The Mussel Cookbook

Drawings by Edith Allard

The Mussel Cookbook

Sarah Hurlburt

Harvard University Press
Cambridge, Massachusetts
London, England

Library of Congress Cataloging in Publication Data
Hurlburt, Sarah, 1925–
    The mussel cookbook.

    Includes index.
    1.   Cookery (Mussels)   I.   Title.
TX753.H87   641.6'94   76-49634
ISBN 0-674-59542-4

For Graham, Sarah, Beau,
Cyndy, and Leeds

Preface

This book is devoted to a virtually new seafood for Americans, the wonderful and versatile *Mytilus edulis,* commonly called the edible blue mussel. In Europe, surely 95 percent of the population is familiar with this delicious meaty mollusk—and has been for centuries. In North America today, an equally large percentage of the population has never heard of edible mussels. Yet along our coasts are thousands of acres of these beautiful shellfish growing wild, virtually unnoticed. Because of the immense popularity of mussels in Europe they are commercially farmed, and still the demand exceeds the supply. Until this happens in the United States and Canada (and it surely will), readers of this book will have a head start on providing their families, friends, and customers with an inexpensive, nutritious, and delectable new seafood.

Americans are already familiar with clams, oysters, shrimp, and other seafoods. These are very expensive and will become more so, as they are being overharvested. Yet mussels are more abundant, more easily gathered, equally tasty and nutritious, and can be prepared in a greater variety of ways. Mussels are delicious opened fresh and eaten on the half shell like oysters or littleneck and cherrystone clams. They can be broiled, steamed, or prepared in over a hundred different ways; they can be substituted for clams in any recipe. Cooked mussels taste like a subtle blend of oysters and clams; mussel meats are sweeter than clams. Also, mussels can be frozen, canned, pickled, and smoked.

My own love affair with mussels began about twenty years ago when the Hurlburt family packed up and left my native North Carolina and went north to New England, which was my husband's home and which has been our home ever since. Because of Graham's love of the out-of-doors and the sea, I was quickly introduced to all sorts of strange plants, game birds, animals, and edibles from the ocean. These creatures were plunked down in the kitchen with the indication that they were good to eat (most were) and that it would be nice if they could be prepared for supper.

As my husband never claimed more than marginal cooking ability (although he has been involved with food all his life and is a graduate of the Cornell University School of Hotel Administration), it became my task and pleasure to experiment with cleaning and preparing these wild delicacies.

Early on I became acquainted with mussels, and they soon rated among my favorites—so abundant, so delicious, and so unappreciated by Americans. I quickly learned how adaptable they were to a wide variety of cooking methods. Still, most people—even seaside residents—were reluctant to try mussels, even though they could seldom give a good reason for their presumed dislike. So initially when I served mussels to friends, I would disguise them in some simple or exotic dish. After the meal, when my guests voiced their pleasure and it was revealed that they had just enjoyed their second serving of mussels, they would invariably say, "I don't believe it; they really were delicious. Could I have your recipe?"

When the opportunity came to travel abroad to countries where mussels have long been enjoyed as daily fare, I knew that I wanted to collect and adapt local recipes and write a book extolling the true virtues of this marvelous mollusk. This volume is the result of my efforts, and it is my hope that it will bring an exciting new food to the North American table.

From the day when man first placed a piece of meat on the end of a stick and held it over a fire until the present time with its countless methods of cooking technology, I suspect that each advance has been merely an improvement or variation on the work of others. With that thought I acknowledge my debt to cooks and technicians back to the caveman.

More important, I should like to thank the people from whom I have solicited recipes and advice and those who have heard of my endeavors and given me their favorites. In particular, I am grateful to Craig Claiborne and Pierre Franey of the *New York Times*, who sampled some of my recipes at our home in Duxbury, Massachusetts. And my special appreciation goes to Eileen

O'Brien, whose organizational help and cheerful adaptability helped this book become a reality.

Last, and most important of all, I should like to say thank you to my family for their enthusiastic support. Without their help I should not have been able to write *The Mussel Cookbook.*

S.H.

Contents

Foreword

This enlightening volume on mussel cookery provides an in-depth account of a food so little known to people in the United States that most tables of food composition do not even list it. Sarah Hurlburt describes the characteristics of mussels, their habitats, their nutritive value, their history as a foodstuff, and the techniques used to farm them. She writes of her own introduction to them and of the degree of their availability on today's market. They taste, she tells us, like "a subtle blend of oysters and clams." Thus informed about the nature and desirability of mussels as a food, the reader is given detailed information on basic cooking techniques and an abundance of appetizing recipes.

Mussels, which are mollusks similar in nature to oysters and clams, are an excellent source of high-quality protein. Like most seafoods, they are low in fat, particularly saturated fat, when compared with other protein sources such as beef, pork, or lamb. And they are good sources of minerals—calcium, phosphorus, iron, and the so-called trace minerals. You need not be concerned about the cholesterol content of mussels, or for that matter of other shellfish. A few years ago it was discovered that the methodology long used for measuring cholesterol in foods was in error when applied to shellfish. When this was corrected, the cholesterol content of shellfish was found to be comparable to that of other fish. Just go easy on the butter you may add, or—better—use a poly-unsaturated margarine! When a prudent diet is the order of the day, mussels are quite suitable for inclusion on the menu.

By stimulating interest in this "new" food, Mrs. Hurlburt has led us to better knowledge of a largely untapped food supply at a time when the use of all food resources is increasingly important to feed our expanding population.

Fredrick J. Stare, M.D.
Professor of Nutrition
Harvard University School of Public Health

Introduction

Cawdel of Muskels

*Take and seeth muskels, pyke hem clene
and waishe hem clene in wyne. Take almandes
and bray hem. Take sime of the muskels and
grynde hem, and some hewe small. Draw
the muskels yground with the self broth.
Wyrny the almandes with faire water.
Do all this togider. Do thereto verjous and
vinegar. Take whyte of lekes, and parboil hem
wel. Wryng out the water, and hew hem
small. Cast oile thereto, with onyons parboiled, and minced small. Do thereto powder,
fort, safron, and salt; a lytel seeth it, not
to stondyng, and messe it forth.*

—A 1390 English recipe attributed to the Master of
the Cooks of King Richard II

The elegant beauty of a single mussel.

Why Mussels?

The edible blue mussel is a dark blue or brownish bivalve mollusk, closely related to the clam and the oyster. Abundant along both coasts of the Atlantic and Pacific, the mussel is well adapted to withstand strong currents and surf; it thrives in the intertidal zone, the area between the high- and low-tide lines, where other shellfish often cannot live.

One of the major secrets of this success is the mussel's ability to attach itself to almost anything with its byssus threads, a bundle of fine but tough brown fibers also called the beard. Firmly anchored to a rock, a pier, or other mussels, it can withstand heavy storm seas. In fact, mussels have even been used to support a stone bridge where the tide was so rapid that it wore away the mortar faster than it could be repaired. Mussels were put in the spaces between stones and attached themselves so firmly that their byssus threads actually held the bridge together.

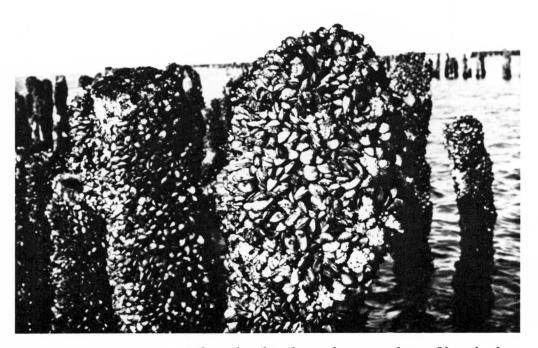

Clumps of mussels attached to pilings by means of their byssus threads. They will fasten in the same manner on virtually anything with which they come in contact.

Like other bivalves, the mussel is a filter feeder. Its protective shells are tightly closed if exposed at low tide, but when covered with water they are held slightly open. Water is drawn into the animal by the action of cilia, microscopic hairlike projections from cells in the gills. Plankton (microscopic plants and animals) and detritus (minute organic particles) are filtered out of the stream of water and transported into the mouth. The current of water swept through a mussel is more powerful than that through any other mollusk.

Filter feeders depend on the quality and purity of the water they filter and the nutrients in that

4   *Introduction*

water. They tend to concentrate material found in the water, so shellfish from polluted water do not provide wholesome food for other animals, including humans. Certain one-celled plants are perfectly good food for mussels but are highly toxic to larger animals that may eat mussels. These are the algae responsible for the so-called red tides that occur under certain conditions. To protect shellfish-eaters, European, Canadian, and American governments support programs that monitor both the water in shellfish areas and the mollusks themselves. (For more information, see pages 10 and 21.)

Enormous quantities of water pass through every mussel: an adult about 3 inches long filters as much as 10 to 15 gallons in a 24-hour period. It has been calculated that there are thousands to millions of diatoms, one kind of planktonic plant, in each quart of seawater. The apparently infinite amount and detritus of plankton provide an almost inexhaustible supply of primary food, which is rapidly and efficiently converted by mussels into excellent flesh for human consumption.

Mussels are certainly one of the most efficient producers of edible flesh. It has been estimated that for each step an animal is removed from primary plant food, up to 90 percent of the food energy is lost. Filter feeders like mussels are very close to the beginning of the food web, while most commonly eaten fish are carnivorous; big fish have eaten little fish, which have eaten smaller fish, and so on—each time with a large loss of energy.

Beef cattle consume 21 pounds of protein in order to produce a single pound of meat protein.

The meat of the mussel is extremely nutritious. Ounce for ounce, as the chart below shows, mussel meat has about the same amount of protein as beefsteak, much less fat, 25 percent of the calories, and many more mineral nutrients.

No shellfish is substantially more nutritious than mussels, and no other shellfish gives such high yield of meat. Before spawning in spring or summer, about 50 percent of the wet weight is flesh, all of which is edible. Tender, delicious, and easy to digest, mussel meat could be—now and in the future—one of the cheapest, most highly nutritious, and most abundant meats on the world market.

3.5 ounces (*100 grams*) raw meat:	Common blue mussel	T-bone steak (choice)
Calories	95	395
Protein	14.4 grams	14.7 grams
Fat	2.2 grams	37.1 grams
Carbohydrates	3.3 grams	0
Calcium	88 milligrams	8 milligrams
Phosphorus	236 milligrams	135 milligrams
Iron	3.4 milligrams	2.2 milligrams
Thiamin	0.16 milligram	0.06 milligram
Riboflavin	0.21 milligram	0.13 milligram

SOURCE: United States Department of Agriculture Handbook No. 8, *Composition of Foods*, December 1963.

Introduction

Life of the Blue Mussel

Mytilus edulis, the edible blue mussel, is found (and eaten) throughout the world along the coasts in temperate areas. It can tolerate a wide range of salinity and is often abundant in estuaries, where fresh water meets the ocean. A great deal has been written about European mussels, from Norway south into the Mediterranean, and especially about commercial mussel farming (see page 12). The blue mussel is abundant along the eastern coast of Canada and the United States from the Arctic Ocean south to Cape Hatteras, North Carolina, and along the west coast from Alaska to San Francisco; the California mussel *(Mytilus californianus),* a closely related edible species, is common along the west coast of the United States and Mexico. There is also information in the literature about the use of mussels for food in South America and the Orient, including China, Japan, Korea, and the Philippine Islands.

Other species of mussels are edible and are popular in some areas; for example, the horse mussel *(Modiolus modiolus)* makes delicious chowder. The blue mussel, however, is the most tender and has the most delicate scent and flavor. The recipes in this book have been developed specifically for the edible blue mussel.

Blue mussels are 2 to 4½ inches long, 1 to 2 inches high, and ¾ to 2 inches broad (roughly *5 to 11, 2½ to 5, and 2 to 5 centimeters*). The broadest area is near the narrow, pointed end of the shells, which then taper toward the larger, rounded end. Usually somewhat triangular, with the hinge along the short side of the triangle, the two shells are equal in size and shape. The color can range from light blue to purple to almost black to brown. The inside lining of these shells is pearly and iridescent, suitable for many decorative uses. Unlike clams, mussels do not bury themselves in sand and so do not tend to accumulate grit inside the shells.

Just inside each shell lies the mantle, a glandular fold of skin that covers most of the body. The mantle manufactures the shell as the mussel grows. It also extracts oxygen from the seawater passing through the mussel. The gills act as filters to trap food and transport it to the mouth. Two pairs of long smooth bands of tissue, the labial palps, help convey food into the mouth; they probably are sensitive to different odors in the water as well. Connected to the intestine is a large digestive gland, sometimes called the liver. The heart circulates a clear, colorless blood through small blood vessels. There is a simple nervous system.

Although the blue mussel spends most, if not all, of its adult life firmly anchored in one place, it retains the ability to move. The long, muscular foot can be extended outside the shells and attached to an object by means of a sucker at the tip. Muscles of the foot then contract, drawing the body toward it. Young mussels can actually creep up a perpendicular glass wall by this method. The

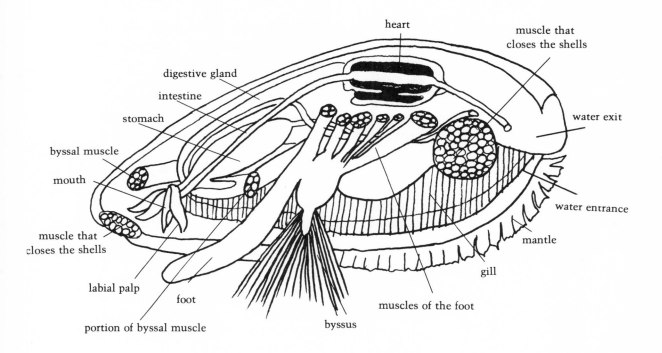

heart

muscle that
closes the shells

digestive gland

intestine

stomach

byssal muscle

mouth

muscle that
closes the shells

labial palp

foot

portion of byssal muscle

byssus

muscles of the foot

gill

mantle

water entrance

water exit

foot also guides the attachment of the byssus
threads, which are secreted by a gland at the base
of the foot.

Reproductive tissue occupies a large part of the
mussel's body before spawning; it colors the
mantle reddish in females and creamy in males.
Water temperature may be an important factor in
determining when mussels spawn. The female releases from 5 to 25 million eggs at one time, and
the male releases a continuous stream of sperm-

containing milt, which causes the water to look milky for several feet. The eggs are fertilized in the water and development is extremely rapid. In a few days free-swimming larvae are formed, maturing to the adult form in a matter of weeks. Young mussels may remain afloat for a month or more before finding a permanent place of attachment; they hang from the surface film by a byssus thread or hold a bubble of gas between the shells. Sexual maturity is reached in about a year, and blue mussels have been known to live ten years or more.

History in America

In Europe information on the history of mussels abounds; in North America, on the other hand, such material is relatively scarce. It is well known that the American Indians consumed mussels, as their shells frequently have been found in unearthed Indian shell heaps.

Mussels were a familiar food for the Pilgrims when they landed in Plymouth and surely helped sustain them over the first few desperate years. *Mourt's Relation* cites Governor William Bradford as saying in 1622: "This bay is a most hopeful place . . . an abundance of muscels, the greatest and best we ever saw." But from the 1620s to the early 1900s little information concerning mussels is to be found.

In 1914 the U. S. government attempted to publicize the value of mussels in the Boston area. The

Bureau of Fisheries launched a daily campaign, with which the newspapers enthusiastically cooperated. Posters and booklets including recipes were distributed to fish dealers. The French chef of a leading Boston hotel promoted the gourmet uses of mussels in elegant restaurants, hotels, and clubs, which were provided with free mussels by the bureau. To introduce mussels to those who did not eat at expensive restaurants, the bureau furnished a barrel of mussels to each police station for the members of the force to take home. Walking their beats the next day, Boston patrolmen spread the word about the delicious new seafood they had sampled. Public lectures, accompanied by mussel dinners, also reached out to the general public.

This campaign was so imaginative and successful that it was reported on several occasions in the *New York Times*. A survey the following year showed that mussels had become a regular item in the popular diet. Worried that the natural mussel beds in the area were in danger of being depleted by the heavy demand for edible mussels, the bureau discontinued its publicity efforts while looking for new sources.

At about the same time mussels of excellent quality were being canned and pickled, but little attempt was made to market them. For the next twenty years mussels again were relatively unknown to the general public, although Americans of French, English, Spanish, Portuguese, and Italian heritage continued to eat them.

Between World War I and World War II efforts of any sort to encourage the use of mussels were rare. But during World War II the wild mussel population was again put under pressure to ease the shortage of local food stocks. Much of this harvest was canned. The government sponsored a further campaign to reintroduce this unfamiliar food to the public. Mussels served in a variety of ways—steamed, fried, cooked in chowder, as raw mussel cocktail—were enthusiastically received by a considerable number of people. People who liked other shellfish rated mussels as excellent.

If mussels have been so successful as emergency rations in times of shortage, why have they not become part of the regular American diet? Americans have long been familiar with mussels as excellent fish bait, as a source of shells to be made into a variety of decorative items, and as producers of mussel mud—a rich, odorous fertilizer. Many Americans may not realize that mussels are edible and delicious, or they may not know how to prepare them. There has also been speculation that some Americans may be prejudiced against eating mussels because of fears that they might be poisonous. This possibility is discussed in some detail further on in the introduction.

Farming Mussels

If mussels are to become a routine part of the American diet, a continuing, reliable supply of high-quality mussels must be available without depleting the natural population. This can be accomplished by mariculture—farming or cultivating the sea. Is a cultivated mussel better than a wild one? Probably yes, although they taste similar. Under controlled conditions they grow faster, they can be harvested when they are fattest, and much higher production can be achieved from a limited area of ocean. Cultured mussels are much more homogeneous than wild ones. Also, mariculture is a profitable industry—the fishermen-farmers make money, as well as the dealers, processors, retailers, and restaurants.

Mussels could be farmed in North America just as they are in Europe, both in areas where they grow wild and in many other waters where they are not now abundant. In Europe they are grown in several different ways, depending on the geography of the coast and the velocity and heights of the tides. There are a number of variations, but the three principal methods are the raft culture of Spain, the pole culture of France, and the bottom culture of the Netherlands. These methods could be adapted to American coastal areas.

The Raft Culture of Spain

On the northwest Atlantic coast of Spain at about the 42nd parallel lie the five Galician bays or rias. They extend far inland and the slope of the land to the water is reasonably steep, not unlike the coast of Maine and eastern Canada. They are protected from the full force of the ocean by islands at their mouths. Mussel raft culture in this area is only about 30 years old. The rafts are rather simple devices. The first ones were made from the hulls of old fishing vessels. Later structures have from 4 to 6 concrete or steel floats or pontoons, and a few new ones are constructed of styrofoam and fiber glass. The size of the rafts varies, but an average one might support 700 ropes, each about 30 feet long, suspended from the rafts. In the spring the young mussel seeds collect directly on bare ropes hanging from the rafts for that purpose. The mussels mature to about 3½ or 4 inches in a year. These are the largest and commercially the fastest grown mussels in the world. The yield of meat to total live weight in the shell is 35 to 50 percent. The mussels have to be thinned and transplanted several times, otherwise their fast growth and bulk would cause them to fall off the ropes.

A 30-foot rope produces over 250 pounds of mussels annually and a 700-rope raft produces over 90 tons of mussels in the shell—as much as 90,000 pounds of drained meat annually. An acre of water surface can support 3 to 5 rafts. In an intensively cultivated area, one acre can produce more than a quarter of a million pounds of meat per year.

Spain is the world's largest producer of cultured mussels. The annual yield is in excess of 220,000 tons in the shell. Ninety-five percent of this is derived from the five Galician bays, where there are over 3,000 rafts. About 45 percent is canned, 5 percent frozen, and the remainder sold fresh.

About one-fourth of the fresh mussels are exported to France and Italy. Although it requires much hard work on the part of entire families, the Spanish mussel mariculture industry is innovative, uncomplicated, and profitable.

A fisherman wrapping a rope with seed mussels. When he has finished, he will suspend the rope from the raft on which he is working. This routine is standard practice among mussel farmers of the Galician bays in Spain.

The Pole Culture of France

From the south Atlantic coast of France, north through the coastal regions of Brittany and Normandy, mussel mariculture is accomplished by the pole or bouchot method. Here the coast is made up of long, gradually sloping beaches extending far out to sea. The ocean floor is unprotected from storms and other vagaries of nature. The tremendous tides are perhaps the most unusual feature of much of this coastline. In some northern areas there is at times a difference of 50 feet between high and low water. Oak poles about 8 inches in diameter are driven into the ocean floor in long rows approximately 3 feet apart, with 12 feet or so between rows. The top 5 feet of the poles, exposed at low tide, is where the culture takes place.

The pole method was accidentally discovered in the mid-thirteenth century. The principle involved has remained virtually unchanged since that time. An Irish sailor named Walton was shipwrecked and put ashore at Esnandes near La Rochelle in southwest France. When he sank some poles into the ocean mud and stretched nets between them to catch sea birds for food, he quickly observed that mussels in great abundance grew on the poles. Thus began the bouchot system of mussel mariculture. In the Bay of Aiguillon where Walton came ashore, there are now over 2½ million poles—more than 50,000 rows of 50 poles each. Along the French coastline today, there are about 700 miles of these rows of mussel poles.

Fishermen who work the bouchots by foot, ox cart, tractor, or little flatbottom mud boats called açon—miles out on the flats near Mont-Saint-Michel—must be ever conscious of getting back to the mainland before the tremendous tidal bore surges toward them. As in Spain, the government of France leases the mussel-growing areas to the

farmers, and most of the bouchots are conducted as a family enterprise. A single pole will yield 20 to 25 pounds of live mussels per year, or about 10 pounds of meat. One acre will yield about 5 tons of live mussels, or over 4,000 pounds of meat annually. France produces over 50,000 tons of live mussels annually, all sold fresh. In fact, the demand for live mussels is so great that France imports almost as many mussels as are grown locally.

M. André Bouyé, director of the Mussel Growers Association of France, is shown working his bouchots from a small boat at Charron. He is thinning the mussels, which will then be transplanted to other poles for further growth.

The Bottom Culture of the Netherlands

This third method of farming the sea is closest to growth of wild mussels, but the harvesting,

A harvest of mussels being unloaded from fishing boat to dockside processing area. This procedure typifies the bottom-culture method used in North Sea countries.

cleaning, and storage are more highly mechanized. Mussel farming has existed for more than 300 years in the Netherlands on the bottom of the shallow, partially diked or enclosed seas. The mussel farmers here also lease their culture plots from the government.

The wild mussel seed is dredged up by boat from natural growth areas when it is 1/3 to 1/2 inch long. It is then transplanted to culture plots at depths of 10 to 20 feet. These seed mussels mature to their marketable size of approximately 3 inches in about 20 months.

The Dutch mussel growers currently produce in excess of 100,000 tons of mussels annually, over 30,000 tons of meat. Eighty percent of the production is sold fresh, most exported to France and Belgium; the remainder is processed into cans and jars for worldwide distribution. With the bottom culture method the Dutch farmers can produce approximately 25 tons of live mussels per acre annually, or about 15,000 pounds of meat per acre.

These methods of mussel mariculture have evolved into large, profitable industries, actually creating sizable quantities of high-quality meat and significant employment. One acre of grazing land can produce 300 pounds of clear boneless beef. Compare that to 4,000 pounds of clear mussel meat in France, 15,000 pounds in the Netherlands, and up to 300,000 pounds in Spain. Imagine 43 billion pounds of meat per year from an area of water the size of Cape Cod Bay, which is less than 300 square miles! At the annual mussel production rates now being achieved in northwest Spain, this is theoretically possible. This much high-protein meat would provide every living person in the United States with a ton of meat per year. It is technically feasible now.

Where to Buy

Locating fresh mussels at the present time may take a bit of detective work in some areas. If fish retailers do not stock them, they may be able to fill special orders with some advance notice. Often you can buy relatively small amounts directly from seafood wholesalers. French, Italian, Spanish, Portuguese, Greek, Armenian, or Turkish restaurants that serve mussels can provide information, as well as delicious examples of mussel cookery. Check with grocery stores serving these ethnic communities. Calls to newspaper food editors, gourmet shops and delicatessens, home economists, cooking teachers, or government agencies (national, state, and local) may be useful. You may discover small noncommercial sources simply by asking people in areas near mussel beds; if you do obtain mussels this way, read the next section on gathering mussels that are safe.

If you can locate canned or bottled mussels more readily than fresh, you can always use them in recipes that call for mussel meats obtained by steaming fresh mussels. However, canned or bottled clam juice—rather than the brine of the mussels—should be substituted for mussel broth. Mussels packed in oil, pickled, smoked, or prepared with a sauce cannot be substituted in recipes.

Canned mussels soon should be readily available through supermarkets and specialty stores. And I feel certain that in the relatively near future frozen mussels will be found on our supermarket shelves in a variety of forms and package sizes. These will include mussels that are block frozen, individually quick frozen, on the half shell, breaded for frying, and the principal ingredient in a number of frozen entrées and convenience-food items. More and more restaurants will feature mussels fried, steamed, in chowders, en brochette, and in many of the ways presented in this book. With this inevitable introduction of prepared mussels by restaurant and supermarket chains, people in the Midwest will become as familiar with the new seafood as those who live in Boston or Brussels.

Gathering Mussels That Are Safe to Eat

If you are near a source of wild mussels, you may want to spend a delightful hour or two gathering the meat for your dinner on the beach or rocky shore. Before you begin to harvest these fruits of the sea, however, be sure to check with the local or county health officers or shellfish wardens about any permits or restrictions, and especially about possible health hazards.

For centuries people have known that illness and even death occasionally resulted from eating shellfish. Many superstitions arose about how to avoid toxic shellfish. More recently, scientific studies have identified the causes of such poisonings: sewage or chemical pollutants in the water can be concentrated in the meat of filter feeders. Paralytic shellfish poisoning (which used to be called mussel poisoning) was recorded as early as 1793 on the coast of British Columbia.

In fact, it is a one-celled dinoflagellate, eaten by mussels, that actually contains the so-called paralytic shellfish toxin. These tiny organisms of the genus *Gonyaulax* are probably present in small numbers all the time. When local conditions favor their growth, they can rapidly reach concentrations as high as a million organisms in each quart of seawater. Such blooms discolor areas of water, which appear red or brown and are called red tides. Mussels, some kinds of clams, and other filter feeders ingest *Gonyaulax* without becoming ill from the toxin. However, animals that eat mollusks with high levels of this toxin may be badly poisoned. Once the *Gonyaulax* population declines, the shellfish gradually eliminate the toxin from their flesh and are again safe to eat.

Water contaminated with bacteria, viruses, or toxic chemicals may not appear polluted. Even a temporary sewage leak or overflow can render filter-feeding bivalves unsafe as food for days or weeks. European, Canadian, and American governments therefore monitor levels of the poisonous red tide organisms and implement quarantines on harvesting mussels and some kinds of clams whenever necessary. Shellfish that do not meet safety standards may not be harvested or sold, and affected waters are closed to shellfishing until the mollusks have purged themselves of the toxin.

Shutting down of affected mussel beds may not severely hamper large commercial shellfish concerns, which can obtain mussels from many locations, but mussels in your local area may not be safe for a large part of the summer. Times that red tides are likely to occur vary from place to place and year to year. California, for example, quarantines mussels from May 1 to October 31 every year; it is reassuring to know that there have been no instances of paralytic shellfish poisoning during the nonquarantine period. In the northeastern United States, where red tides are still infrequent, the only fatal poisonings have been from uninformed harvesting in officially closed areas. To my knowledge, no commercial mussels or clams have resulted in illness. Some towns post signs when an area is closed to shellfishing for safety reasons, but *you cannot assume that any unposted area is safe.* Always contact the appropriate authorities for up-to-date information.

Since you cannot test the mussels you gather, as a large commercial dealer is required to, it is a wise precaution to know the early signs of paralytic shellfish poisoning. A feeling of numbness in the lips and tingling or burning of the fingers or toes are the first indications. Since this numbness in some cases can become progressive, vomiting should be induced and medical assistance obtained immediately.

After you have determined that conditions are safe, assemble your equipment: a bucket, wire basket, or burlap bag; a screwdriver if mussels will have to be pried off rocks; and gloves to protect your hands if the water is cold. The best mussels can be picked from the water's edge at low tide; they have spent most of their lives under water and will be larger and fatter than mussels that have been exposed to the air twice a day. Mussels can be gathered from sand, mud, rocks, or pilings.

The author and her husband gathering mussels at low tide in Duxbury Bay, Massachusetts.

Don't take more mussels than you plan to use. If you are going to use them within a day or two, be selective and pick just the ones you want—2½ to 3 inches long. However, if you plan to wait more than a couple of days before using them, pick them in clumps, the way they naturally grow. Rinse the clumps briefly in the water and shake them gently. This way they will not be injured by disturbing their byssus threads and will remain alive and fresh in a plastic bag in the refrigerator for up to a week.

Cleaning a mussel by scraping with a dull knife.

Preparing the Mussels for Use

Cleaning

If you harvest your own mussels, give them a few shakes in the salt water to get rid of most of the mud, but that's all. If you buy them fresh at the store, leave them as they are until you are ready to use them. The less you disturb them—by shaking, scrubbing, pulling apart, rinsing, and the like—the longer they will stay alive and fresh. Just before you use them is the time to clean them.

Separate the mussels by pulling them apart. Discard the debris: broken shells, weeds, and any dead mussels. Wash the mussels quickly and thoroughly in running water until all signs of mud are gone. You may want to scrub them individually with a stiff brush or dull knife, thus ridding them of barnacles and the like. Again, eliminate any that might be dead.

Checking for aliveness by attempting to force the two half shells apart.

There are several ways to check to see if a mussel is alive and fresh. After it has been in the air undisturbed for a while, it may open slightly; it should close when you tap it. Hold the mussel between your thumb and forefinger and try to force the two half shells apart by pressing and sliding; it should resist stubbornly. When tapped with a knife, the mussel should sound solid, not hollow. Finally, mussels that have been washed should smell fresh and sweet like the ocean, not strong like decayed fish.

Now that the mussels are clean and you have assured yourself that they are alive, remove the byssus threads or beard. These threads emerge from between the two shells on the concave side.

Removing the byssus threads by means of a sharp pull.

It is these threads that are used by the mussels to attach themselves to each other, as well as rocks, pilings, and almost anything else. At the far end of each thread is a little foot or suction cup. While the threads are edible, they are tough and often full of dirt; it is best to remove them. With scissors, cut the byssus threads off as close to the shell as possible. Or you can remove even more of them if you give them a good yank—pulling from the small pointed end of the shell toward the larger round end. You may even pull out a small amount of meat, but this is quite all right.

With mussels most of the dirt is on the outside and not the inside. Soft-shell (steamer) clams, for instance, accumulate sand and mud inside their shells, which must be purged before eating. Mussels do not succumb to this contamination as a rule.

Opening Raw Mussels

Opening mussels to get the fresh meat is simple. It is easier than opening oysters or clams, and you should not end up with broken shells or cut fingers.

Clean the mussels and remove the byssus threads. Hold the mussel between your thumb and forefinger with the large round end of the mussel fitting into the curve of your hand. The convex side of the mussel should be against the inside of your forefinger. Insert the blade of a small, sharp, thin knife between the shells about a third of the way down from the small pointed end. Draw the knife through the mussel completely until it emerges through the large round end.

Now spread the two shells open—they will remain hinged. Unlike the oyster, which generally has one flat shell and one rounded shell, mussel shells are both rounded and almost identical. By splitting the two shells apart with a knife, you have also cut the raw meat in half. Work your knife around each shell carefully, loosening the meat in each half so it will drop out. If the mussels are very fresh, there will be a good bit of liquor in each mussel. Two dozen fat fresh mussels 2¹/₂ to 3 inches long will yield about a cup of drained mussel meats and a scant cup of liquor. After you have opened a dozen or so, you will be skilled and able to open them neatly and quickly (perhaps 4 per minute).

You may prefer to remove the meat whole (like an oyster on the half shell). While this procedure

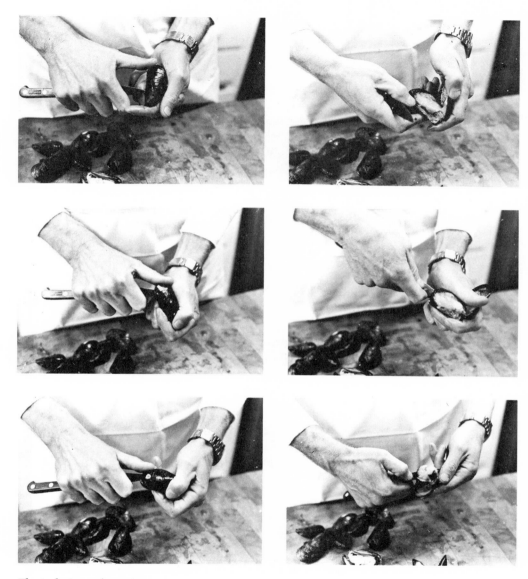

The technique of opening a mussel, showing the point of insertion and path of the knife, the spreading apart at the hinge, and the removal of the meat.

The handsome result: mussels on the half shell!

requires a little more dexterity, it is not at all difficult. Hold the mussel as described above and insert the point of the knife in the same place. But instead of pulling the knife blade straight through the meat, tilt the point upward and work it around the upper shell. This is like working a knife around the inside edge of a glass held upside down, or freeing a soft-boiled egg from its shell. When you have worked the blade all around the shell, open the two shells with your fingers and discard the top one. Then with the blade, free the meat from the bottom shell; it will be in one piece for eating raw or further preparation.

A microwave oven can be used to open mussels easily and successfully. Set the control on "high" (if you have a variable-temperature oven), place the mussels on a shallow glass casserole, and put it in the oven. For mussels of average size, 3 minutes' microwave exposure will yield almost the same product as if you had opened the mussels raw. A setting of 3½ to 4 minutes produces meats similar to those that have been steamed in the conventional manner. The broth will remain in the bottom of the dish to be used when called for in your recipe.

Steaming to Remove the Meat

Put cleaned mussels with the byssus threads removed in a heavy pan, pot, or kettle on the stove. Do not steam too many mussels at one time. I usually fill the pot two-thirds full, making sure the mussels are no deeper than 8 inches. For a larger amount, use a second pan. If the mussels are very fresh (24 hours maximum) you do not have to add any liquid. Otherwise add a half-cup of water or white wine for each 2 quarts of mussels.

Cover the pot and let steam from 5 to 7 minutes at high heat, until the shells are open wide and the meats are just firming and becoming loose from the shells. To continue steaming after they have opened will only toughen the meats.

Remove the mussels from the broth, and when cool remove the meats from the shells. (Save the shells if you will need them later.) If any byssus threads are left, they may be removed at this point. Also some people, for dishes requiring an especially tender product, will remove the dark "elastic" band from around the circumference of the meat. I do not usually bother to do this.

The cloudy broth in the bottom of the kettle is delicious. Strain it through several layers of cheesecloth or paper coffee filters to remove any grit or sediment.

If you want smaller mussel meats, you can slice them with a very sharp knife or cut them with sharp scissors. Mussels are tender and soft, so unless your recipe says specifically to do so, do not put the meats through a food chopper or blender. You will get mush instead of small pieces.

Preserving

Canning is not considered the safest way to preserve shellfish in the home kitchen. The long processing yields a product inferior in flavor and texture and lacking in virtually all the desirable fresh qualities. However, mussel canning is possible; information on how to proceed may be obtained from the Bureau of Fisheries, U. S. Department of the Interior, Washington, D. C.

The best method of preserving mussels is freezing; for ideal results they should be prepared and frozen immediately after gathering and steaming open. If the mussels are frozen in broth (as in the last recipe of the **Soups and Stews** section), canning jars, plastic containers, or clean milk or ice cream cartons may be used. For freezing without liquid, plastic bags are best, since they can be snugly packed with little or no air remaining. Shellfish must be quick-frozen at an initial temperature of −20° F *(−29° C)*, then may be stored for 3 to 4 months at 0° F *(−18° C)*.

Frozen mussels may be cooked while still frozen unless they are to be breaded or coated with batter for frying. For chowder, stews, or cream-sauce dishes they may be used in the frozen state. If you thaw them, do not refreeze.

Estimating Yield

As in all living things, individual size and weight of mussels vary. In the approximations given, an average length of 2³/₄ inches *(7 centimeters)* is assumed.

1 bushel of fresh mussels weighs 60 pounds *(27 kilograms)*
1 bushel of fresh mussels contains 1,000 mussels
1 quart *(1 liter)* **of fresh mussels weighs 1¹/₂ pounds** *(750 grams)*
1 quart *(1 liter)* **of fresh mussels contains 25 mussels**
3 pounds *(1¹/₂ kilograms)* **of fresh mussels when steamed will produce approximately 1 pound** *(500 grams)* **of mussel meats**
1 quart *(1 liter)* **of fresh mussels when steamed yields 1 cup** *(¹/₄ liter)* **of mussel meats**
1 cup *(¹/₄ liter)* **of steamed mussel meats weighs about ¹/₂ pound** *(250 grams)*
1 cup *(¹/₄ liter)* **of drained mussel meats contains about 25 meats**

In compiling the recipes for this cookbook, it has seemed wise to take cognizance of the worldwide conversion to the metric system that is currently under way. Where appropriate, ingredient quantities are given in both American and metric (except for teaspoons and tablespoons, which can be used in both systems) with the thought that at some future date, only the metric will be required. Mussel cooks will be in the vanguard, as they will have already served their apprenticeship!

Quantity Preparation

As any experienced chef, restauranteur, or cook knows, mussels (like other shellfish) are best prepared "to order" or in small quantities.

Many of these recipes, however, can be utilized directly in, or easily adapted to, commercial or banquet kitchens. Dishes such as steamed mussels, chowders, soups, and casseroles may be expanded into quantity recipes by straight multiplication. Seasonings, spices, and garnishes may be adjusted to meet individual preferences. The recipes particularly recommended for quantity preparation are marked with a *[Q]*.

Simple Dishes to Enjoy Indoors or Out

There is no flavor more delectable than the freshness and sweetness of mussels just gathered and steamed in a large kettle by the dunes. Or, if you wish to return home after your day at the beach, you can feast in a matter of minutes by means of these basic methods.

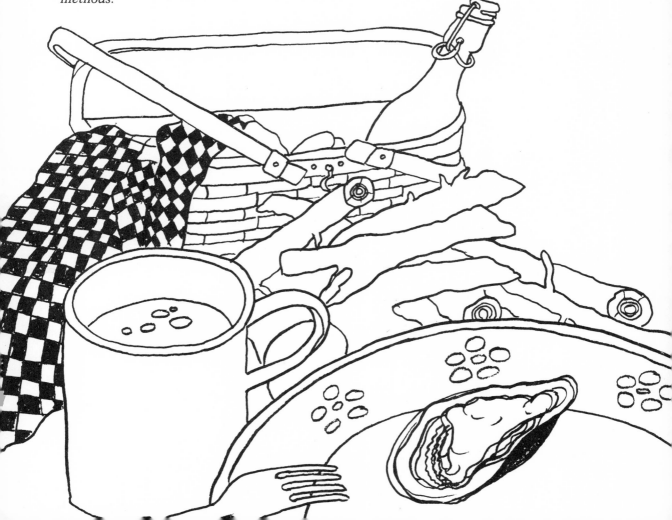

Picnic on the Beach

Gather about 1 quart (1 liter) of mussels, or about 25, for each person. Bring along at least ⅛ pound (65 grams) butter or margarine per person.

For cooking and serving, you need a hot driftwood fire, a bucket or kettle with cover for steaming, a small pan for melting butter, and cups for sipping broth.

Wash mussels in clean seawater to remove all sand. Check whether they are alive, then beard them (see full cleaning instructions on page 25).

Put mussels in kettle, to a depth of no more than 8 inches *(20 centimeters).* (For a larger amount cook another batch.) Do not add water. Cover kettle and place over fire. Put butter in pan alongside to melt. Shake kettle a few times to redistribute contents and make sure all cook evenly. Mussels should steam open in 5 to 7 minutes and be ready to eat. Discard any that do not open.

Dip hot mussels into butter. Sip the fragrant natural broth.

On the Half Shell

Mussels are delicious raw—plain or with the same accompaniments used for raw oysters.

You must be very sure that those you eat raw are alive when opened and from uncontaminated waters. Read all information in the introduction about gathering mussels that are safe to eat or about checking wholesomeness of those purchased.

For each appetizer serving allow 8 to 10 mussels with shells, at least 2³/₄ inches *(7 centimeters)* long. Clean them, check aliveness, and open according to instructions on page 28.

At the beach most enjoy personally gathered raw mussels out of hand from the shell, adorned with nothing more than lemon juice (if that).

For a special picnic or appetizer at home, arrange opened mussels in shells on a bed of crushed ice and garnish with lemon wedges. A tomato seafood cocktail

Simple Dishes to Enjoy Indoors or Out

sauce is good with them; a recipe is given with **Batter-Fried Tidbits** on page 61. Thin sandwiches of brown bread and butter, a traditional oyster accompaniment, are a perfect match for the rich flavor of mussels.

Wherever you enjoy mussels raw, open them just before serving.

Éclade (Wilderness Mussels on a Board)

4 quarts *(4 liters)* **uncooked mussels in their shells, cleaned and shells well scrubbed according to instructions on page 25**
French bread and fresh sweet butter

This is one of the oldest French ways—and perhaps the best—to prepare mussels. They are arranged, still in the shell, on a board, then covered with dry pine needles. The needles are set afire to cook the mussels open and give them a delicious flavor. A sophisticated bottle of cold Muscadet makes a fine complement to the simple natural delicacy of the mussels.

You will need a large wooden board about 1¹/₂ by 2 feet *(45 by 60 centimeters),* or 2 smaller ones of equivalent measurement; several nails at least 2 inches *(5 centimeters)* long and a hammer; and tinder-dry pine needles, enough to cover sides and top of board with a layer 4 inches *(10 centimeters)* deep.

Arrange board in sand on beach taking care that sparks from needles cannot possibly ignite brush or forest. Drain and dry mussels. Place them side by side with pointed ends upward so that ashes will not get inside when shells open. The arrangement will be a big rosette design, a circle of shell "petals."

The tricky part is propping up the first 4 mussels. If you are a beginner, you may want to drive a nail or two in the center of the board to lean these against. Skilled cooks do without nails and set the 4 mussels in a cross or square shape so that they support one another.

When all are arranged, cover sides and top with needles and set afire. When needles burn out, fan ashes away. Eat mussels at once with bread and butter.

On the Grill

4 quarts *(4 liters)* **uncooked mussels in their shells, cleaned and shells well scrubbed according to instructions on page 25**
Melted butter or margarine
Lemon wedges

Arrange a grill 4 to 5 inches *(10 to 12¹/₂ centimeters)* above intense coals of a wood or charcoal fire. Place mussels in shells on grill (on flat pan if necessary) and cook until most of them open. Discard those that do not open. Serve at once with melted butter to dip in and lemon to squeeze over.

Oven-Broiled Variation

Put mussels on broiler pan about 5 inches *(12¹/₂ centimeters)* below preheated broiler until they open. Watch closely to avoid overcooking, which will dry and toughen the meats.

Wine Pot

Mussels steamed in wine can be prepared on the beach from those you have just gathered or at home from those previously gathered or purchased.

At the beach you will perhaps want to allow 1 quart (1 liter) of mussels per person. For a home meal each quart could serve two people as a main dish with accompaniments.

Steam mussels open, using wine as liquid, according to instructions on page 31. Serve hot in shells. Strain broth through several layers of cheesecloth or paper coffee filters and serve in individual cups, or add unmelted butter to hot broth to melt. Serve lemon wedges to squeeze over mussels and/or into broth.

3 quarts *(3 liters)* **uncooked mussels in their shells, cleaned and shells well scrubbed according to instructions on page 25**
¹/₂ cup *(1 deciliter)* **dry white wine**
Butter or margarine
Lemon wedges

Beer Bowl

Serves 4

1 can or bottle (12 ounces, or 3½ *deciliters*) **beer**
1 **small onion, chopped**
1 **tablespoon lemon juice**
1 **bay leaf, crushed**
4 **quarts** *(4 liters)* **uncooked mussels in their shells, cleaned and shells well scrubbed according to instructions on page 25**
2 **tablespoons butter or margarine**
Hot boiled potatoes

Prepare potatoes in advance. Put beer, onion, lemon juice, bay leaf, and mussels in large kettle. Cover and steam on high heat for 5 to 7 minutes, or until most shells open. Discard those that do not open.

Distribute mussels into soup bowls. Strain broth through several layers of cheesecloth or paper coffee filters into saucepan. Add butter and simmer uncovered for about 3 minutes to reduce slightly; pour over mussels. Serve with hot potatoes.

44

Outer Beach Bake

Serves 8

You will need a large kettle (at least 5-gallon, or 2-decaliter, capacity) with lid for this dish, which may be cooked indoors or out.

Put ingredients in kettle in order listed. Cover and steam over low heat for about 45 minutes, or until chicken is tender and sausage is fully cooked. Serve hot broth in cups with the "bake." Do not eat mussels that have not opened.

8 medium-sized unpeeled potatoes for boiling
8 small onions, peeled
4 quarts *(4 liters)* **uncooked mussels in their shells, cleaned and shells well scrubbed according to instructions on page 25**
8 chicken legs
12 link sausages
8 ears corn, shucked
4 cups *(1 liter)* **water**

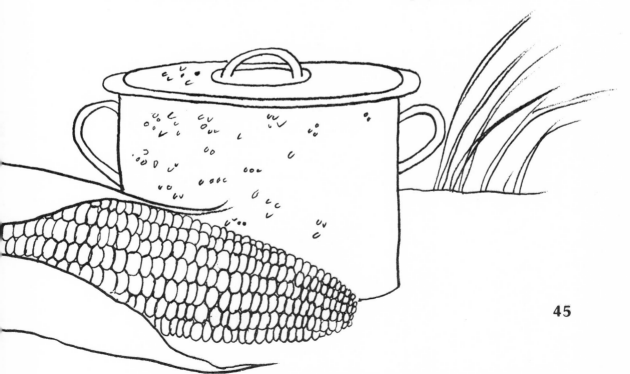

A Dozen Quick-and-Easy Tricks

Toothpick Appetizer   Marinate cooked
mussels in olive oil and lemon juice with one
or all of the following: minced green onion,
minced parsley, or dill seed.

Quick Dip   Combine ½ cup *(125 grams)*
chopped cooked mussel meats with 1 cup
(2½ deciliters) sour cream, 1 teaspoon dehydrated minced onion, and 1 tablespoon bacon
bits. Chill and serve with chips or crackers.

Easy First Course   To mussels on the half
shell, either fresh or cooked, add a rounded
teaspoon frozen spinach soufflé that has been
thawed. Top with bacon bits, broil 5 minutes
at 3 inches *(7½ centimeters)* below heat, and
take straight to the table.

An Oldie   Try the familiar put-whatever-it-is-
on-top-of-a-deviled-egg routine with chopped
cooked mussel meats.

Lunch on the Boat   To a can of clam bisque,
add about a dozen cooked mussel meats,
1 tablespoon butter, and 1 teaspoon
Worcestershire sauce. Salt and pepper to
taste. Heat and serve over toast in soup
plates; garnish with paprika.

Salad Savvy   Toss cold cooked mussels into a
green salad with fresh tomatoes. This is the
most frequent use for them in Sweden, according to a Swedish friend.

Simple Dishes to Enjoy Indoors or Out

Remoulade   A quite acceptable hurry-up version can be created by combining 1 cup *(250 grams)* cooked mussel meats, ¹/₂ cup *(60 grams)* chopped celery, ³/₄ cup *(2 deciliters)* mayonnaise, and ¹/₂ cup *(1 deciliter)* chili sauce. Chill and serve on crisp lettuce. Also good as sandwich filling on dark rye or brown bread.

Skewers Orientale   Thread a few bamboo skewers with cooked mussel meats, dip in teriyaki or barbecue sauce, and broil on hibachi.

Rarebit Tout de Suite   Add ¹/₂ cup *(125 grams)* cooked mussel meats to a can or package of frozen welsh rarebit. Heat as directed on container and serve over toast. Top with strips of fried bacon.

In-a-Hurry Hash   Combine 2 cups *(500 grams)* cooked mussel meats with 2 cups *(250 grams)* diced cooked potatoes and 2 tablespoons dehydrated chopped onion. Salt and pepper to taste. Brown in ¹/₄ cup *(¹/₂ deciliter)* shortening over medium-high heat for about 15 minutes, turning frequently.

Instant Paella   Prepare a box of Spanish rice mix according to directions on package. Combine 2 cups *(500 grams)* cooked mussel meats and 1 rounded tablespoon minced green pepper, and add to rice. Pour into casserole and bake 15 minutes in preheated 350° F *(180° C)* oven. If mixture seems too dry, add a little tomato juice.

Simple Dishes to Enjoy Indoors or Out   **47**

Casserole a la Casa   Cover the bottom of a small well-buttered casserole with 2 layers of cooked and drained mussel meats. Top with a can of undiluted cream of mushroom soup. Sprinkle 2 tablespoons sherry over all, and bake in preheated 350° F *(180° C)* oven for 25 minutes.

Appetizers, Many Doubling as Main Dishes

There is an endless variety of ways in which mussels may be served as a first course. Many resemble other familiar seafood appetizers—shrimp cocktail, marinated herring, and oysters or clams on the half shell. Portions should always be small, interestingly seasoned, and attractive in appearance.

Cream Cheese Dip

Makes about 3 cups [Q]

Combine cream cheese and sour cream in blender; mix in remaining ingredients with fork. For best flavor chill 2 hours, but serve at room temperature. Accompany with crackers, chips, or melba toast.

3-ounce *(90-gram)* **package softened cream cheese**
1¹/₂ cups *(3¹/₂ deciliters)* **sour cream**
1 cup *(250 grams)* **cooked, chopped, and drained mussel meats (canned, or fresh steamed by method on page 31)**
2 tablespoons finely chopped onion
1 tablespoon finely chopped parsley
1 teaspoon capers
Dash of hot-pepper sauce
Crackers, chips, or melba toast

Spicy Cocktail

Serves 4 [Q]

¹/₄ **cup** *(65 grams)* **butter or margarine**

2 **cups** *(500 grams)* **cooked and drained mussel meats (canned, or fresh steamed by method on page 31)**

3 **onions, chopped**

2 **cups** *(5 deciliters)* **strained mussel broth saved from steaming, or canned or bottled clam juice**

3 **tablespoons white vinegar**

2 **bay leaves**

2 **whole cloves**

Salt

1³/₄ **cup** *(3¹/₂ deciliters)* **mayonnaise**

²/₃ **cup** *(1¹/₂ deciliters)* **chili sauce**

Lemon juice

Lemon slices or wedges for garnish

Melt butter in saucepan over low heat; add mussels and onions. Sauté until light brown. Stir in broth, vinegar, bay leaves, and cloves. Simmer 30 minutes on very low heat. Taste and add salt as necessary. Cool and refrigerate.

For sauce, blend mayonnaise and chili sauce with lemon juice to taste; chill.

To serve, arrange mussels on platter around bowl of sauce; add lemon garnish.

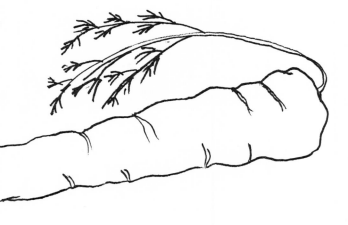

Appetizers, Many Doubling as Main Dishes

Marinated Mussels with Sour Cream

Serves 6 [Q]

In a 2-quart *(2-liter)* crock or other non-metallic container with cover, mix all ingredients except lettuce and sour cream. Refrigerate 24 hours.

To serve, drain mussels from marinade. Place on lettuce-lined salad plates; put a dollop of sour cream on top of each. Garnish with vegetables from the marinade.

2 cups *(500 grams)* **cooked and drained mussel meats (canned, or fresh steamed by method on page 31)**
1 cup *(2½ deciliters)* **strained mussel broth saved from steaming, or canned or bottled clam juice**
1 carrot, peeled and sliced
2 onions, thinly sliced
2 cloves garlic, minced
1/8 teaspoon ground allspice
1/4 teaspoon crumbled basil
1/4 teaspoon crumbled tarragon
1 bay leaf
Freshly ground black pepper
Pinch of cayenne
1 teaspoon salt
1 cup *(2½ deciliters)* white vinegar
2 tablespoons olive oil
Lettuce leaves
Sour cream

Aspic-Stuffed Shells

Serves 6

36 fresh mussels in their shells, at least 2³/₄ inches *(7 centimeters)* long, steamed by method on page 31

1 cup *(2¹/₂ deciliters)* mussel broth saved from steaming, strained through cheesecloth or coffee filters

1 envelope (¹/₄ ounce, or *8 grams*) unflavored gelatin softened in ¹/₄ cup *(¹/₂ deciliter)* cold water

1 cup *(2¹/₂ deciliters)* dry white wine

¹/₂ cup *(130 grams)* minced celery

1 teaspoon lemon juice

Several dashes cayenne or hot-pepper sauce

Garnishes: 36 pimiento strips, each about 1¹/₂ inches *(3³/₄ centimeters)* long; finely chopped dill pickle

Remove mussel meats and discard half the shells; lay remaining shells side by side on baking sheet. (If you chill pan of shells, final gelling will be hastened.) Finely chop mussels and drain.

In saucepan, combine broth and softened gelatin; heat, stirring, until gelatin is fully dissolved. Remove from heat; add mussels, wine, celery, lemon juice, and cayenne. Pour into shallow pan and refrigerate just until very thick but not fully gelled; do not let it become firm.

Stir mixture gently to divide ingredients evenly; spoon into shells. Decorate each with pimiento strip; refrigerate until completely gelled. Garnish servings with chopped pickle.

Mussels
à l'Escargot

Here mussels are prepared with garlic-herb butter of the type used for escargots (snails). This appetizer may be assembled days ahead, frozen, and popped in the oven at the last minute.

Remove mussels and discard half the shells. Blend butter and remaining ingredients thoroughly. Put a dab in each remaining shell, lay a mussel on it, and cover with more butter mixture.

Put shells side by side in shallow baking pan. Bake in preheated oven at 400° F *(200° C)* for 15 to 20 minutes, until very hot and bubbly. Serve on individual plates with cocktail forks, or on a tray with small picks. Appetizers should be sizzling hot.

24 fresh mussels in their shells at least 2³/₄ inches *(7 centimeters)* **long, steamed by method on page 31**
¹/₂ cup (¹/₄ pound, or *125 grams***) softened butter or margarine**
1 tablespoon finely chopped parsley
1 tablespoon lemon juice
1 shallot, minced
3 large cloves garlic, puréed
1 slice uncooked bacon, minced
Salt and pepper to taste

Bacon-Broiled Bits

Serves 4 [Q]

24 cooked whole mussel
  meats (canned, or fresh
  steamed by method on page
  31)
Lemon juice
1 clove garlic, puréed
About 3 tablespoons soy sauce
12 slices bacon, each cut in
  half
About ¼ cup *(½ deciliter)*
  brandy (optional)

Pat mussels dry with paper toweling and season with lemon juice, garlic, and soy sauce. Marinate a few minutes if you like.

Drain mussels and wrap each in a bacon piece; secure with wooden toothpick. In large frying pan over medium-high heat, sauté until bacon is crisp; drain on paper towels and serve sizzling hot. Appetizers also may be cooked under broiler or in oven. Arrange apart on baking pan—or, even better, on wire rack in baking pan. Turn once while broiling. Or bake in preheated 450° F *(230° C)* oven until bacon is crisp.

If, as a gourmet touch, you want to flame these with brandy, drain all fat from frying pan or use chafing dish at table. Warm appetizers in pan and brandy in saucepan. Pour brandy over appetizers, tip pan, and touch match to edge to ignite.

Mussels à la Grecque

Serves 4 to 6 [Q]

Steam mussels open according to instructions on page 31, adding lemon, onion, garlic, and bay leaf to steaming pan. Remove and discard one shell from each mussel; put mussels in remaining shells side by side in shallow baking pans that can be placed under broiler.

Melt butter in frying pan; add mushrooms and sauté until liquid has mostly disappeared. Add olives and parsley; sauté until lightly browned, seasoning with salt and pepper. Put a little of this mixture on top each mussel.

Broil about 5 inches *(12½ centimeters)* below preheated broiler for about 15 minutes, or until topping is bubbly and mussels are hot.

2 quarts *(2 liters)* **uncooked mussels in their shells, cleaned and shells well scrubbed according to instructions on page 25**
1 lemon, sliced
1 onion, quartered
2 cloves garlic, minced or puréed
1 bay leaf
3 tablespoons butter or margarine
¾ cup *(45 grams)* sliced fresh mushrooms or 4 ounces *(125 grams)* canned sliced mushrooms, drained
½ cup *(80 grams)* sliced black olives
1 tablespoon parsley
Salt and pepper to taste

Grilled Mouclade

2 tablespoons butter or margarine
2 quarts *(2 liters)* **uncooked mussels in their shells, cleaned and shells well scrubbed according to instructions on page 25**
$^1/_2$ **cup** *(1 deciliter)* **dry white wine**
2 shallots or green onions, finely chopped
$^1/_2$ **cup** *(60 grams)* **finely chopped celery**
1 bay leaf
Pinch of crumbled thyme
$^1/_2$ **cup** *(1 deciliter)* **heavy (whipping) cream**
Suggested accompaniment: buttered toast

Melt butter in large saucepan; pour into shallow baking dish. To saucepan add mussels, wine, shallots, celery, bay leaf, and thyme. Cover and steam over high heat 5 to 7 minutes, shaking pan occasionally, until shells open. Remove and discard shells; put mussel meats in baking dish.

Simmer broth to reduce to half. Remove bay leaf and add cream; pour over mussels. Place about 4 inches *(10 centimeters)* under preheated broiler for 5 to 8 minutes, or until hot and bubbly (do not let brown). Serve very hot with buttered toast.

Singapore Pupus

Serves 8

In Hawaii appetizers are called pupus. *Many are Oriental in character, as this one is.*

In bowl, combine all ingredients except crumbs and oil. Mix well, then form into balls about 1 inch *(2½ centimeters)* in diameter; roll each in bread crumbs. Heat oil to 375° F *(190° C)* and fry a few at a time until golden brown. Serve piping hot with small picks inserted. Makes about 24 appetizers.

2 cups *(500 grams)* cooked, chopped, and drained mussel meats (canned, or fresh steamed by method on page 31)
2 green onions, finely chopped
½ cup *(100 grams)* finely chopped and drained water chestnuts
2 tablespoons chopped pimiento
1 teaspoon soy sauce
¼ teaspoon ground ginger
1 tablespoon sherry
1 egg, well beaten
1 cup *(120 grams)* fine dry bread crumbs
Salad oil for deep-frying

Batter-Fried Tidbits

Serves 6

2 cups *(500 grams)* **cooked
and drained whole mussel
meats (canned, or fresh
steamed by method on page
31)**
3 tablespoons lemon juice
3 tablespoons olive oil
1 tablespoon minced parsley
¹/₄ teaspoon crumbled basil
Freshly ground black pepper
Batter (recipe follows)
Salad oil for deep-frying
**Seafood or tartar sauce (recipe
for Tomato Seafood Sauce
follows)**

2 tablespoons olive oil
2 cups *(5 deciliters)* **lukewarm
water**
³/₄ teaspoon salt
2¹/₂ cups *(350 grams)* **flour**
3 egg whites

*Partially make the batter ahead—the day
before or several hours before frying. Add
beaten egg whites just before cooking.*

Marinate mussels at least 30 minutes in
the lemon juice, olive oil, parsley, basil,
and pepper to taste.

At cooking time pat mussels dry a bit
on paper toweling so that batter will ad-
here. Dip each mussel in batter; fry a few
at a time in oil heated to 375° F *(190° C)*,
until golden brown. Serve with sauce.

Batter

Mix oil, water, and salt, then fold flour in
gently to prevent elastic texture. Let
batter rest at least 1 hour. Just before
cooking time, beat egg whites until they
hold firm peaks and fold into batter.

Tomato Seafood Sauce

Combine all ingredients. Makes 1½ cups (3½ deciliters).

1 **cup** (2½ deciliters) **catsup**
½ **cup** (1 deciliter) **chili sauce**
1 **tablespoon vinegar**
1 **teaspoon Worcestershire sauce**
1 **teaspoon prepared horse-radish**
Juice of 1 lemon
¼ **teaspoon crumbled basil**
4 **or 5 drops hot-pepper sauce**

Stuffed Mushrooms

Serves 6 [Q]

36 fresh mushrooms, at least
  2 inches *(5 centimeters)* in
  diameter
1 large clove garlic, chopped
6 tablespoons butter or margarine
1 medium-sized onion, finely
  chopped
1½ cups *(180 grams)* fine dry
  bread crumbs
1 cup *(250 grams)* cooked,
  chopped, and drained
  mussel meats (canned, or
  fresh steamed by method on
  page 31)
3 tablespoons sherry
¼ teaspoon finely crumbled
  oregano
Salt and pepper to taste

Wash mushrooms and pat dry on paper toweling. Remove stems and finely chop half of them (use remainder another time in a stew or an omelet).

Sauté the garlic in butter about 1 minute or until soft; toss in chopped stems and onion; sauté until golden. Remove from heat; add bread crumbs, mussels, sherry, and oregano. Mix well, taste, and add salt and pepper.

Salt mushroom caps inside, then spoon in stuffing. Butter baking pan and put mushrooms on it, stuffing side down. Preheat broiler and broil mushrooms about 4 inches *(10 centimeters)* below heat for about 2 minutes, or until hot and beginning to soften. With spatula carefully turn each over and broil until light golden brown. Serve piping hot.

Appetizer Piquant

*Serves 10 to 12 as appetizer,
6 as main dish [Q]*

In saucepan, sauté onion in oil until golden. Add pickles, parsley, capers, and catsup; remove from heat and cover.

At serving time steam mussels until they open by method on page 31. Heat serving plates in oven.

Remove and discard one shell from each mussel and put mussels with shells intact on warm plates. Add wine to sauce and bring to a simmer, stirring constantly; do not boil. Immediately spoon over mussels and serve hot. Or the dish may be served chilled.

2 tablespoons finely chopped onion
¹/₂ cup *(1 deciliter)* **olive oil**
¹/₄ cup *(30 grams)* **finely chopped sour pickles**
1 tablespoon chopped parsley
1 tablespoon chopped capers
¹/₄ cup *(¹/₂ deciliter)* **catsup**
4 quarts *(4 liters)* **fresh mussels in their shells, cleaned and shells well scrubbed according to instructions on page 25**
2 tablespoons dry white wine

Mussels Casino

**36 fresh mussels in their
shells, at least 2¹/₄ inches
(7 centimeters) long,
steamed by method on
page 31**
Rock salt
¹/₂ green pepper, chopped
**2 tablespoons finely chopped
onion**
**2 tablespoons chopped
pimiento**
**1 teaspoon prepared horse-
radish**
Dash of hot-pepper sauce
**3 tablespoons softened butter
or margarine**
**3 strips lean bacon, each cut
in 1-inch** *(2¹/₂-centimeter)*
squares
1 lemon, cut in 6 wedges
Parsley sprigs for garnish

Remove mussels and discard half the
shells. Spread layer of rock salt in large
shallow baking pan suitable for taking
to the table or in 6 individual baking
dishes. Set mussels, each in a shell, in
salt.

Blend green pepper, onion, pimiento,
horseradish, and hot-pepper sauce into
butter. Put dollop of butter mixture and
then a bacon piece on each mussel.
Broil 4 inches *(10 centimeters)* below
preheated broiler for 6 to 8 minutes, or
until bacon is crisp. Garnish with lemon
and parsley.

Appetizers, Many Doubling as Main Dishes

Moules à la Table

Serves 2 to 4 [Q]

This recipe makes 2 large or 4 small appetizer servings. To serve more people, increase ingredient amounts as necessary, but do not cook more than 24 mussels at a time. Prepare additional servings in batches. This is easy to do because cooking time for each batch is only about 5 minutes. You can do the cooking at the table in a chafing dish or an electric frying pan.

Put oil, onion, parsley, and mussels in large frying pan or chafing dish. Cook over high heat, shaking pan back and forth and dribbling in wine little by little as mussels begin to open. As soon as most have opened (discard any that do not), pour gently into soup plates. Serve with French bread to dip into broth.

1 tablespoon olive or other vegetable oil
2 tablespoons minced onion
2 tablespoons finely chopped parsley
24 uncooked mussels in their shells, cleaned and shells well scrubbed according to instructions on page 25
2 cups *(5 deciliters)* or ½ bottle dry white wine
French bread

Mushrooms Rissolés

Serves 6

½ cup (¼ **pound, or** *125 grams*) **butter or margarine**
¼ **pound** *(250 grams)* **fresh mushrooms, finely chopped**
1 **cup** *(250 grams)* **cooked, chopped, and drained mussel meats (canned, or fresh steamed by method on page 31)**
2 **tablespoons chopped parsley**
1 **tablespoon grated onion**
1 **tablespoon lemon juice**
Pinch of cayenne
Salt to taste
Pie-crust dough (enough for a 2-crust pie)
Salad oil for deep-frying

For filling, melt butter in frying pan, add mushrooms, and sauté until lightly browned. Add mussels, parsley, onion, lemon juice, and cayenne. Mix and add salt to taste; cool.

Roll out dough ⅛ inch *(3 millimeters)* thick and cut into 2½-inch *(6-centimeter)* circles (makes approximately 18). Put about 1 tablespoon filling to one side of circle, brush edges with water, fold dough over, and pinch edges to seal turnover.

Heat oil to 375° F *(190° C)*; fry a few turnovers at a time until golden brown. Serve hot. (*Note:* Turnovers may be frozen and reheated from the frozen state in a hot oven.)

Appetizers, Many Doubling as Main Dishes

Mussels Rockefeller

Serves 4

Fill 2 glass pie plates (9-inch, or *22½-centimeter*, size) with rock salt to depth of ½ inch *(13 millimeters)* and sprinkle with water. Remove and discard one shell from each mussel.

   Melt butter in frying pan. Stir in onion, celery, and parsley; cook over low heat until onion is tender, about 4 minutes. Remove from heat; add spinach and remaining ingredients. Mix well. Spoon about 1 tablespoon of mixture over each mussel in half shell. Place on rock salt and bake in preheated 450° F *(230° C)* oven for 10 minutes. Serve immediately.

Rock salt
24 **large fresh mussels (steamed by method on page 31)**
4 **tablespoons butter or margarine**
2 **tablespoons finely chopped onion**
2 **small stalks celery, minced**
2 **tablespoons chopped parsley**
1 **cup** *(180 grams)* **cooked, chopped spinach, well drained**
¼ **teaspoon ground thyme**
2 **dashes hot-pepper sauce**
½ **cup** *(60 grams)* **fine dry bread crumbs**
¼ **teaspoon salt**
1 **tablespoon lemon juice**

Soups and Stews, Both Light and Main Dish

Mussels lend themselves delightfully to soups and stews, which have been a basic food of every country and every civilization through the ages. Clear soups are considered, by and large, to be a first course. Thick soups may be used as a main course and are unusually high in nutrition and low in calories. These resemble chowders, soups, and stews with which you are already familiar—clam chowder, oyster stew, and lobster bisque, for instance.

Mussel Broth

Serves 4 [Q]

Combine broth and herb in saucepan.
Cover and simmer until fully flavored,
about 10 minutes; salt if necessary.

Serve hot with lemon slice in each cup.
Make several long shreds of carrot to garnish, using potato peeler or coarsest
blades of shredder.

4 cups *(1 liter)* **strained
mussel broth saved
from steaming**
**1 teaspoon dried thyme or
chervil**
Salt to taste
4 thin slices lemon
Raw carrot, peeled

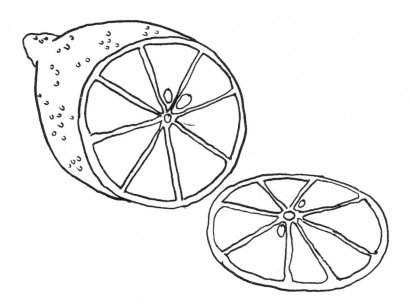

Mussels with Lemon

¹/₂ cup (¹/₄ pound, or *125 grams*) **butter or margarine**
1 **large carrot, peeled and chopped**
1 **tablespoon minced green onion**
1 **clove garlic, minced**
3 **lemons, very thinly sliced**
1 **bay leaf, crushed**
2 **tablespoons chopped parsley**
Salt and pepper to taste
2 **quarts** *(2 liters)* **uncooked mussels in their shells, cleaned and shells well scrubbed according to instructions on page 25**
3 **tablespoons flour**
French bread

A waterfront café in Marseille specializes in this colorful way of serving mussels. A checked tablecloth and pottery bowls will make you feel as if you were there.

In large kettle, melt half the butter; add carrot, onion, and garlic. Sauté a minute or two, or until onion is soft. Add lemon slices, bay leaf, parsley, salt, pepper, and mussels.

Cover and cook rapidly over high heat 5 to 7 minutes, or just until mussels open. Discard any that do not open. Remove mussels and keep warm in covered pan in oven. Strain broth and reserve it along with some of the lemon slices.

In a saucepan, melt remaining butter and blend in flour, stirring until it foams. Remove from heat and add broth. Cook, stirring constantly, until thickened.

Distribute mussels and lemon slices in soup plates; pour sauce over. Serve with French bread.

Chinese-Style Soup

Serves 4 [Q]

Bring chicken broth to a boil in large pot. Add mussels and enough water to cover. Add garlic, oil, and pepper. Cover and steam until shells open, about 5 to 7 minutes. Discard any that do not open. Remove mussel meats from opened shells and discard shells. Strain broth through several layers of cheesecloth or paper coffee filters.

Put mussel meats into warmed soup bowls and cover with hot broth. Garnish with onion slices and some of the tops chopped fine. Noodles or rice may be served on the side or put in the soup.

3 cups *(7½ deciliters)* **salted chicken broth**
2 quarts *(2 liters)* **uncooked mussels in their shells, cleaned and shells well scrubbed according to instructions on page 25**
Water
2 cloves garlic, crushed
2 tablespoons peanut or other oil
Black or crushed red pepper
2 green onions, thinly sliced
Cooked noodles or rice

Mussels
Prague

Serves 6 to 8 as soup course, 4 as main dish

3 quarts *(3 liters)* **uncooked mussels in their shells, cleaned and shells well scrubbed according to instructions on page 25**
3 or 4 tablespoons butter or margarine
2 cups *(300 grams)* **finely chopped onion**
1/2 cup *(50 grams)* **chopped parsley**
4 cloves garlic, minced
1 heaping teaspoon paprika
3 cups *(7 1/2 deciliters)* **water**
1 teaspoon bottled beef extract or 2 beef bouillon cubes
Salt to taste
French bread

This recipe from Czechoslovakia is the favorite of one of my friends.

Combine all ingredients except salt and French bread, cover, and simmer until mussels open; remove and discard any that do not open. Cover and simmer 1 hour more on lowest heat, adding salt as needed. Serve in soup plates with French bread for dunking.

Soups and Stews, Both Light and Main Dish

Quick Stew
for Two

Toasted water biscuits or sodá crackers are a recommended accompaniment. This recipe will serve four as a soup course.

Remove mussel meats from their shells and discard shells. Melt butter in saucepan. Add mussel meats and sauté 1 minute. Slowly stir in milk and cream. Heat, but do not boil. Season to taste with salt, pepper, and mace or nutmeg.

About 24 uncooked mussels in their shells, cleaned and opened raw by method on page 28
¹/₄ **cup** *(65 grams)* **butter or margarine**
1 **cup** *(2¹/₂ deciliters)* **milk and 1 cup** *(2¹/₂ deciliters)* **cream, or 2 cups** *(5 deciliters)* **half-and-half**
Salt and pepper to taste
Pinch of mace or nutmeg

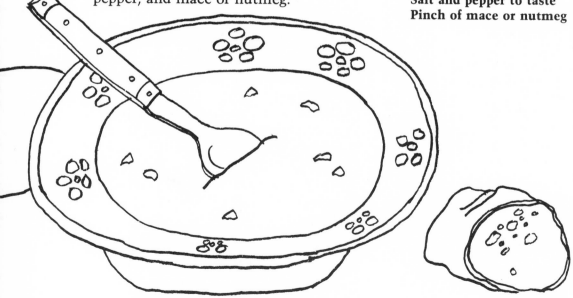

Soups and Stews, Both Light and Main Dish

Down-Maine Stew

*Serves 16 as soup course,
8 as main dish*

4 ounces *(125 grams)* **salt
pork, finely diced**
8 **quarts** *(8 liters)* **uncooked
mussels in their shells,
cleaned and shells well
scrubbed according to in-
structions on page 25**
2 **large onions, thinly sliced**
1½ **cups** *(3½ deciliters)* **water**
Salt and pepper to taste

*Chester Whitaker, a mussel fisherman
from Walpole, Maine, shared his family's
recipe with me.*

In large kettle, fry salt pork until lightly
browned. Remove pork scraps, drain on
paper towel, and reserve. Add mussels,
onion slices, water, and a little salt and
pepper. Cover and bring to a simmer over
high heat; lower heat and simmer until
mussels open, about 8 minutes. Discard
any that do not open. Adjust salt and
pepper seasoning. Serve in the shells with
the broth in soup plates. Garnish with
pork scraps.

(*Note:* If any stew is left over, make a
quick chowder the next night. Discard
mussel shells. To mussel meats and
broth, add cooked diced potatoes and
cream or milk in amount to equal
amount of leftover broth. Heat, but do
not boil.)

Rhode Island
Red Chowder

*Serves 12 as soup course,
6 as main dish [Q]*

*This chowder can be stretched with a
cup or two of water if you have unexpected
guests.*

In large kettle over medium heat, fry salt
pork until lightly browned. Remove
pork scraps, drain on paper towel, and reserve for garnish.

Add onions and sauté until limp but
not yet beginning to brown. Add mussels,
broth, green pepper, potatoes, and bay
leaf. Cover and simmer until potatoes are
almost done. Pour in tomatoes and tomato juice; season with salt and pepper.
Cover and simmer 30 minutes more.
Garnish servings with pork scraps and
serve with biscuits or crackers.

4 ounces *(125 grams)* **salt
pork, finely diced**
**3 medium-sized onions,
chopped**
4 cups *(1 kilogram)* **cooked
mussel meats (canned,
or fresh steamed by method
on page 31); cut large ones
in half**
4 cups *(1 liter)* **strained
mussel broth saved from
steaming, or canned or bottled clam juice**
¹/₂ cup *(80 grams)* **chopped
green pepper**
4 cups *(500 grams)* **uncooked
diced potatoes**
1 bay leaf
1 cup *(250 grams)* **canned
whole tomatoes**
3 cups *(7¹/₂ deciliters)* **tomato
juice or cocktail vegetable
juice**
Salt and pepper to taste
**Toasted water biscuits or un-
salted crackers**

Country Chowder

3 ounces *(100 grams)* **salt
pork, finely diced**
2 teaspoons **flour**
2 cups *(5 deciliters)* **water**
2 cups *(5 deciliters)* **strained
mussel broth saved from
steaming, or canned or bottled clam juice**
1 large **onion, chopped**
3 **potatoes, peeled and diced**
Salt and pepper to taste
2 cups *(500 grams)* **cooked
mussel meats (canned, or
fresh steamed by method on
page 31)**
2 cups *(5 deciliters)* **light
cream or half-and-half**
Dash of nutmeg
**Soda crackers or water bis-
cuits**

*Chaudiére is the name of a large boiler or
pot in which French fishermen cooked their
fish stews in ancient times. Our own chowders
and stews probably sprang from this source.*

In large kettle, fry salt pork until lightly
browned. Remove scraps and drain on
paper towel; save for garnish.

Blend flour into fat, then add water,
broth, onion, potatoes, and a little salt
and pepper. Cover and simmer until potatoes are done. Add mussels, cream, and
nutmeg. Heat, but do not boil; add more
salt and pepper as necessary. Garnish
with pork scraps. Soda crackers or water
biscuits are the usual accompaniment.

Chowder Base to Freeze
Prepare **Country Chowder** without the
cream; freeze. To serve, thaw, pour into
top of double boiler, add cream, and heat
gently without boiling. Add nutmeg and
salt and pepper as needed.

Cream Soup

Serves 6 [Q]

In top of double boiler over hot water, melt butter and blend in flour. Add milk and cream; cook, stirring, until slightly thickened. Add mussels, salt, and mace. Continue cooking and stirring about 10 minutes, or until thoroughly warm.

²/₃ **cup** *(165 grams)* **butter or margarine**
3 **tablespoons flour**
6 **cups** *(1 ½ liters)* **milk**
1 **cup** *(2 ½ deciliters)* **heavy (whipping) cream**
2 **cups** *(500 grams)* **cooked mussel meats (canned, or fresh steamed by method on page 31), coarsely chopped**
³/₄ **teaspoon salt**
Dash of mace or nutmeg

THE SANITARY
BUTTER JAR

Cream

Beautiful Bisque

3 tablespoons butter
2 teaspoons minced green
   onion
1 cup *(250 grams)* **cooked
   mussel meats (canned, or
   fresh steamed by method
   on page 31)**
1 cup *(2¹/₂ deciliters)* **strained
   mussel broth saved from
   steaming, or canned or bottled clam juice**
1 cup *(2¹/₂ deciliters)* **milk
   and 1 cup** *(2¹/₂ deciliters)*
   cream, or 2 cups *(5 deciliters)* **half-and-half**
1 teaspoon lemon juice
2 egg yolks, well beaten
Salt and pepper to taste
Sour cream
Watercress or parsley for garnish, in sprigs or chopped

This creamy soup thickened with egg yolks is marvelous served chilled on a hot summer day. It's also fine hot.

Place butter and onion in top of double boiler over hot water. Cook, stirring, until onion is soft. Add mussels, broth, milk, cream, and lemon juice; heat. Add a little hot soup to the egg yolks and blend immediately; stir slowly into soup. Season with salt and pepper while stirring, and warm just until hot enough to serve. Garnish servings with dollops of sour cream, and watercress or parsley.

Billi Bi Soup

Serves 4 [Q]

This soup was created by the world-renowned Maxim's of Paris in honor of William B. Leeds, an American tin magnate who dined there regularly.

In large kettle, combine mussels, onion, shallots, and wine. Cover and bring to a boil, lower heat, and simmer about 5 minutes or until mussels open. Remove from heat and take out mussels and shells; discard any unopened mussels.

Pour liquid through a strainer lined with cheesecloth or through paper coffee filters. Heat with cream, but do not boil; season to taste with cayenne, salt, and pepper.

Serve either hot or cold from a tureen, garnished with a few mussels in their shells. (Use remaining mussels in another dish.)

2 quarts *(2 liters)* **uncooked mussels in their shells, cleaned and shells well scrubbed according to instructions on page 25**
¹/₄ cup *(35 grams)* **chopped onion**
3 tablespoons chopped shallots
1 cup *(2¹/₂ deciliters)* **dry white wine**
2 cups *(5 deciliters)* **heavy (whipping) cream**
Pinch of cayenne
Salt and pepper to taste

Bouquet de Provence

Serves 4 as soup course,
2 as main dish

6 tablespoons olive oil
3 large cloves garlic, minced
1 small onion, chopped
2 fresh tomatoes, quartered
  and seeded
12 pitted black olives
Pinch of crumbled thyme
2 tablespoons chopped parsley
Salt and pepper
2 quarts *(2 liters)* uncooked
  mussels in their shells,
  cleaned and shells well
  scrubbed according to instructions on page 25
French bread

A hearty and delicious soup that will bring
a flavor of the Mediterranean to your table.

Heat olive oil in large frying pan or heavy
saucepan. Add garlic and onion; sauté
until golden brown. Add tomatoes,
olives, thyme, parsley, and salt and
pepper to taste; stir until hot.

Turn heat to high, add mussels, and
cook until shells open wide, basting
shells frequently with pan juices. Discard
mussels that do not open. Serve in soup
plates with French bread for soaking up
sauce.

Bouillabaise

Serves 6 as main dish

It is said that in ancient times the goddess Venus invented this saffron-flavored fish soup as a treat for her husband, Vulcan. Poets and writers have sung its praises through the years, and no seafood book would be complete without a recipe for bouillabaise.

Pour olive oil into heavy saucepan large enough to contain all ingredients. Add garlic, onion, parsley, and rosemary; sauté over low heat until onion is limp but not yet beginning to brown. Add tomato sauce and wine. Cover and simmer 20 minutes.

Add shrimp, fish, and mussels. Cover and simmer about 10 minutes, just until fish is no longer translucent in center; baste with sauce several times during this cooking. Discard mussels that do not open. Add saffron, season with salt and pepper. Serve in soup plates with lemon garnish.

1/4 **cup** *(1/2 deciliter)* **olive oil**
2 **large cloves garlic, puréed**
1 **medium-sized onion, chopped**
1/2 **cup** *(50 grams)* **chopped parsley**
Pinch of crumbled rosemary leaves
1 **cup** *(2 1/2 deciliters)* **tomato sauce**
1/4 **cup** *(1/2 deciliter)* **dry white wine**
1/2 **pound** *(250 grams)* **uncooked fresh shrimp (shells but not tails removed, dark veins cleaned)**
2 **pounds** *(1 kilogram)* **fish fillet—cod or flounder**
24 **uncooked mussels in their shells, cleaned and shells well scrubbed according to instructions on page 25**
3/4 **teaspoon saffron**
Salt and pepper to taste
Thin lemon slices for garnish

Zuppa di Cozze

48 uncooked mussels in their
shells, cleaned and
shells well scrubbed ac-
cording to instructions
on page 25
½ cup *(1 deciliter)* dry
white wine
2 cloves garlic, minced
¼ cup *(35 grams)*
chopped onion
¼ cup *(40 grams)* diced
unpeeled zucchini
1 carrot, peeled and diced
½ cup *(1 deciliter)* olive oil
½ teaspoon crumbled basil
½ teaspoon freshly ground
black pepper
1 can (about 1 pound, or
500 grams) tomatoes, preferably Italian plum type,
with juices
2 teaspoons grated lemon peel
Pinch of cayenne
Salt to taste
Buttered croutons
1 heaping tablespoon flour
dissolved in a little
cold water
Chopped parsley for garnish

Drain mussels and place in large pan
with wine. Cover and steam 5 minutes;
discard mussels that do not open. Add
remaining ingredients except croutons, flour, and parsley. Cover and
simmer 30 minutes. Prepare croutons
and set aside.

Stir dissolved flour into soup and cook
a few minutes longer to thicken and
let flavors mingle. Pour into soup plates.
Sprinkle with parsley, and serve immediately with croutons.

How to Freeze Fresh Mussels in Wine Broth

Makes 12 cups including broth [Q]

You will need a pan of at least 6-quart *(6-liter)* capacity with lid, cheesecloth, and 6 freezer containers of 1-pint *(5-deciliter)* capacity.

Put full quantity of all ingredients except mussels in pan, then add mussels to a depth of no more than 8 inches *(20 centimeters)*. Cover pan, bring liquid to a boil, and cook about 5 to 7 minutes or until shells open. Discard mussels that do not open; set aside those that do. Repeat till all are cooked.

Remove mussel meats and discard shells (save some for garnishing if you like). Divide mussels into containers. Strain broth through several thicknesses of cheesecloth and pour over mussels, leaving a ½-inch *(13-millimeter)* space at top of container. Cover with lid; label and freeze. Can be stored 3 to 4 months at 0° F *(−18° C)*.

Thaw for use in recipes, or serve heated as a first course in small bowls.

4 cups *(1 liter)* **dry white wine**
2 cups *(5 deciliters)* **water**
2 medium-sized onions, chopped
4 carrots, peeled and chopped
6 stalks celery, sliced
Several sprigs parsley
1 bay leaf
12 quarts *(12 liters)* **uncooked mussels in their shells, cleaned and shells well scrubbed according to instructions on page 25**

Salads as Simply Salad or as a Meal

The salads included in this section may be used as side dishes or as complete light meals. They will add zest to your menus and bring color to your table.

You may feel that using mussels in salads is strange. However, some of these recipes are similar to the familiar lobster salad, or to tomatoes or avocados stuffed with shrimp or crab.

Also, you may find that some of the typical European uses appeal to you. Try adding a few cold mussels to green salads, tossing them with potato salad, or garnishing a tomato with them.

Salade Niçoise

Serves 6 [Q]

Line 6 individual salad plates with lettuce. Arrange foods individually but adjoining in a row or circle: tomato wedges, green pepper, eggs, mussels, and olives. Top each serving with 1 or 2 anchovy fillets. Sprinkle with chopped onion, or add onion to vinaigrette dressing before pouring over.

1 head lettuce, separated into leaves
3 tomatoes, cut in wedges
1 small green pepper, cut in strips
4 hard-boiled eggs, quartered
24 cooked mussel meats (canned, or fresh steamed by method on page 31)
Black olives
Green olives
Anchovy fillets
1 tablespoon chopped green onion or chives
Vinaigrette dressing (see Tomatoes Élégantes on next page for recipe)

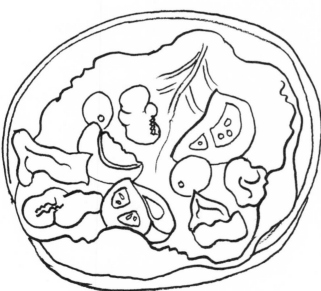

Salads as Simply Salad or as a Meal

Tomatoes Élégantes

2 large ripe tomatoes
Salt and pepper to taste
$^1/_2$ **cup** *(125 grams)* **cooked,
chopped, and drained
mussel meats (canned, or
fresh steamed by method on
page 31)**
1 **cup** *(225 grams)* **small-curd
cottage cheese**
2 **teaspoons chopped chives or
green onion tops (or use cottage cheese with chives
already added)**
$^1/_2$ **cup** *(60 grams)* **chopped
celery**
2 **teaspoons toasted sesame
seeds**
**Loose-leaf lettuce, such as
Boston or Bibb**
2 **black olives**
**Vinaigrette dressing (recipe
follows)**

Cut tops from tomatoes; gently spoon out and discard seeds. Sprinkle with salt and pepper. Turn upside down on paper toweling to drain while you prepare filling.

Combine mussels, cottage cheese, chives, celery, and sesame seeds. Fill tomato shells. Line salad plates with lettuce, place tomato on each, and top each with an olive. Serve with vinaigrette dressing.

Vinaigrette Dressing

Blend dressing ingredients. Mix well before pouring over salad.

1 tablespoon vinegar
¹/₄ cup *(¹/₂ deciliter)* **olive oil**
Salt and pepper to taste
¹/₄ teaspoon prepared mustard,
    preferably Dijon style
Choice of seasonings: minced
    parsley, tarragon, or chervil

Salads as Simply Salad or as a Meal

Stuffed Avocado

2 ripe avocados, halved and
   seeded
Lemon juice
18 cooked mussel meats
   (canned, or fresh steamed by
   method on page 31), each
   cut in 4 pieces
1 cup *(2½ deciliters)* sour
   cream
¼ cup *(30 grams)* chopped
   celery
1 tablespoon chopped chives
   or minced green onion tops
Salt and pepper
Paprika or pimiento strips for
   garnish
Lettuce leaves
Melba toast

Sprinkle avocado halves with lemon juice. Combine mussels, sour cream, celery, and chives; add salt and pepper to taste.

Stuff avocados with mixture and chill. Garnish with paprika or pimiento. Serve on lettuce-lined plates with melba toast.

Yogurt Delight

Serves 4

Mix thoroughly yogurt, vinegar, mint flakes, honey, salt, and pepper. Combine mussels, cucumbers, and chives. Pour yogurt sauce over, and toss gently. Chill and serve on lettuce leaves. Garnish with olives.

1 cup *(2½ deciliters)* **yogurt**
1 teaspoon white vinegar
1 teaspoon mint flakes
½ teaspoon honey or sugar
Salt and pepper to taste
1 cup *(250 grams)* small cooked and drained mussel meats (canned, or fresh steamed by method on page 31)
2 or more cucumbers, peeled and thinly sliced
1 tablespoon chopped chives
Lettuce leaves
8 Greek olives

Fiddler's Green Salad

Serves 4

4 hard-boiled eggs
¹/₂ cup *(1 deciliter)* **mayon-naise**
Dash of hot-pepper sauce
Salt and pepper to taste
1 cup *(250 grams)* **small cooked and drained mussel meats (canned, or fresh steamed by method on page 31)**
2 tablespoons chopped celery
1 teaspoon capers
1 teaspoon chopped gherkins
1 tablespoon chopped parsley
Escarole

On a warm day, this salad is lovely served with melba toast and a glass of iced champagne.

Slice eggs in half lengthwise and remove yolk from whites. Cut egg whites into strips and set aside. With fork, blend until smooth the egg yolks, mayonnaise, hot-pepper sauce, salt, and pepper. Add mussels, celery, capers, gherkins, parsley, and egg-white strips. Stir gently with fork. Chill and serve on bed of escarole.

Salads as Simply Salad or as a Meal

South Sea Salad

Serves 4

Drain pineapple, reserving 4 tablespoons of syrup. Cut 4 pineapple slices in half and set aside. Add reserved syrup to French dressing and marinate remaining pineapple slices for 1 hour. Remove slices and set aside.

Combine mussels, cucumber, celery, and rice. Pour French dressing marinade over and toss gently.

Arrange lettuce leaves in border on round 9-inch *(22-centimeter)* serving dish, filling alternately with rice mixture and grated carrot. Use about half of total rice. Place pineapple half-slices as dividers between rice and carrots.

Shred remaining lettuce, place in center of dish, and top with rest of rice mixture. Decorate with remaining pineapple slices and 3 mussels.

16-ounce *(500-gram)* **can sliced pineapple**
¹/₂ cup *(1 deciliter)* **bottled French dressing**
1 cup *(250 grams)* **small cooked and drained mussel meats (canned, or fresh steamed by method on page 31); set aside 3 for garnish**
¹/₂ cucumber (unpeeled), washed and chopped
4 tablespoons chopped celery
2 cups *(300 grams)* **cooked rice**
1 head Bibb lettuce, half separated into leaves and half left intact
2 medium-sized carrots, peeled and grated

Summer
Mold

Serves 6

1¹/₄ **cups** *(3 deciliters)* **canned
  tomato sauce**
1¹/₂ **tablespoons unflavored
  gelatin softened in** ¹/₄ **cup**
  (¹/₂ deciliter) **cold water**
8-ounce *(250-gram)* **package
  softened cream cheese**
1 cup *(250 grams)* **cooked and
  drained mussel meats
  (canned, or fresh steamed by
  method on page 31), each
  mussel cut in half**
¹/₂ **cup** *(75 grams)* **chopped
  green onion**
1 cup *(2¹/₂ deciliters)* **mayonnaise**
Dash of garlic salt
1 tablespoon lemon juice
Prepared horseradish to taste
Lettuce leaves
Potato chips

Heat tomato sauce, stir in softened gelatin, and beat with wire whisk or electric
mixer until dissolved. Bring to a boil, add
cream cheese, and beat again until
smooth. Remove from heat and cool.

Add mussels, onion, mayonnaise, garlic
salt, and lemon juice; blend. Add horse-
radish to make as zippy as you like.

Pour into ring mold and refrigerate to
gel. Unmold on bed of lettuce. Serve with
chips.

Gourmet Potato Salad

Serves 8 [Q]

The small red potatoes hold their shape better and do not get mushy. Boil them in their skins in salted water and peel while still warm.

In bowl, combine potatoes, mussels, parsley, chives, shallots, and celery seed. In saucepan, combine broth, vinegar, sugar, and mustard; heat just to boiling and pour over salad. Toss, adding salt as necessary. Serve while still warm, garnished with pimiento.

4 cups *(500 grams)* **diced boiled potatoes**
2 cups *(500 grams)* **cooked and drained mussel meats (canned or fresh steamed by method on page 31—reserve broth)**
2 tablespoons chopped parsley
2 tablespoons chopped chives or minced green onion tops
2 tablespoons chopped shallots or green onion
1 teaspoon celery seed
¹/₂ cup *(1 deciliter)* **strained mussel broth saved from steaming, or canned or bottled clam juice**
¹/₂ cup *(1 deciliter)* **white vinegar**
2 teaspoons sugar
¹/₂ teaspoon Dijon-style mustard
Salt to taste
Pimiento strips for garnish

Down-Home, Everyday Main Dishes

*Variations of these dishes are found in most
coastal countries of the world. If you have
some left-over mussels on hand, or a can of
mussels in your cupboard, you can prepare a
delightful lunch or supper with very little
effort.*

Cheese-Chive Omelet

Serves 4

Combine eggs, cream, mussels, cheese, and salt to taste. Warm a serving platter in the oven.

In a large frying pan, melt 2 tablespoons butter over medium heat and sauté chives in it for just a minute. Pour in egg mixture. Lift edges with spatula and tilt pan, letting uncooked egg run to bottom. When top is firm but still moist, fold omelet in half and slide onto warm platter. Sprinkle with paprika and dot with pat of butter.

4 eggs, beaten
¹/₄ cup *(¹/₂ deciliter)* **cream (any type)**
1 cup *(250 grams)* **cooked, chopped, and drained mussel meats (canned, or fresh steamed by method on page 31)**
2 tablespoons grated Gruyère or Swiss cheese
Salt
Butter or margarine
2 teaspoons minced chives
Paprika

Creamed Mussels

3 tablespoons butter or margarine
3 tablespoons flour
2 teaspoons lemon juice
1 teaspoon Worcestershire
   sauce
Hot-pepper sauce to taste
1½ cups *(180 grams)* finely
   chopped celery
2 cups *(5 deciliters)* all-
   purpose cream or half-and-
   half
2 cups *(500 grams)* cooked
   and drained mussel meats
   (canned, or fresh steamed by
   method on page 31)
2 tablespoons white wine or
   sherry
Salt and pepper to taste
Bread for toasting
Watercress or parsley sprigs
   for garnish

Melt butter in top of double boiler over simmering water. Blend in flour, lemon juice, Worcestershire sauce, and hot-pepper sauce. Add celery and sauté until tender; pour in cream slowly, stirring until slightly thickened.

Add mussels, stir until hot; then add wine and salt and pepper to taste. Toast bread and cut each slice diagonally twice to make toast points. Serve mussels rimmed with toast and garnished with watercress or parsley.

Rarebit with Beer

Serves 4 [Q]

Melt butter in top of double boiler over simmering water. Blend in flour, mustard, and cayenne. Gradually stir in beer. Cook, stirring constantly, until smooth and thickened. Blend in cheese; cover and let stand until cheese melts. Beat a little cheese mixture into egg yolk, then gradually stir into rarebit. Cook, stirring, about 2 minutes.

Remove from heat but keep over hot water. Add mussels and salt if needed; cover. Toast bread and cut each slice diagonally twice to make toast points. Serve rarebit rimmed with toast and garnished with paprika or parsley.

1 tablespoon butter or margarine
1 tablespoon flour
1 teaspoon dry mustard
Dash of cayenne
1¼ cups *(3 deciliters)* beer (flat beer generally is used)
1 cup *(100 grams)* firmly packed, shredded sharp cheddar cheese
1 egg yolk, beaten
1 cup *(250 grams)* cooked, chopped, and drained mussel meats (canned, or fresh steamed by method on page 31)
Salt to taste
Bread for toasting
Paprika or parsley sprigs for garnish

Mussels in Tomato Sauce

Serves 4 [Q]

¹/₃ **cup** *(1 deciliter)* **olive oil**
4 cloves garlic, minced
3¹/₂ cups *(800 grams)* **canned tomatoes with juice (chop tomatoes)**
¹/₂ **cup** *(60 grams)* **finely chopped celery**
1 teaspoon sugar
Salt and pepper
4 quarts *(4 liters)* **uncooked mussels in their shells, cleaned and shells well scrubbed according to instructions on page 25**
3 tablespoons chopped parsley

To make sauce, heat olive oil in frying pan and sauté garlic until golden. Add tomatoes, celery, and sugar. Simmer uncovered on low heat, stirring occasionally, until somewhat thickened. Season to taste with salt and pepper.

Steam mussels according to method on page 31. In the oven, warm a wide, shallow baking dish suitable for taking to table. Remove a shell from each mussel and discard. Put mussels in remaining shells side by side in baking dish.

Pour hot sauce over mussels, sprinkle with chopped parsley, and serve at once.

Shellfish
à la King

Melt butter in saucepan and stir in flour. Add mussel broth, stirring constantly.

When sauce boils, stir some of the sauce into the egg yolk, then add to mixture and stir until sauce is thickened. Add mussel meats, pimiento, mushrooms, and almonds. Season to taste and add sherry just before serving.

Serve in warmed patty shells and garnish with sprigs of parsley.

3 tablespoons butter or margarine
3 tablespoons flour
1½ cups *(3½ deciliters)* **strained mussel broth saved from steaming, or canned or bottled clam juice**
1 egg yolk, beaten
2 cups *(500 grams)* **cooked mussel meats, coarsely chopped and drained (canned, or fresh steamed by method on page 31)**
¼ cup *(50 grams)* **pimiento, chopped**
½ cup *(30 grams)* **broiled mushrooms, fresh or canned**
½ cup *(50 grams)* **slivered blanched almonds**
1 ounce *(¼ deciliter)* **sherry**
4 patty shells
Parsley for garnish

Mussels Creole

Serves 6 [Q]

2 tablespoons olive oil
1 medium-sized onion, minced
1 small clove garlic, minced
¹/₄ cup *(40 grams)* **finely chopped green pepper**
1¹/₂ **cups** *(3¹/₂ deciliters)* **canned tomato sauce**
2 medium-sized fresh to-matoes, peeled and diced
¹/₄ cup *(40 grams)* sliced pitted ripe olives
1 teaspoon salt
2 teaspoons sugar
4 cups *(1 kilogram)* cooked and drained mussel meats (canned, or fresh steamed by method on page 31)
¹/₄ cup *(¹/₂ deciliter)* dry sherry
3 tablespoons butter or marga-rine
Hot cooked rice
Chopped parsley for garnish

This dish can be cooked at the table in a chafing dish or an electric frying pan.

Heat oil in frying pan over low heat. Add onion and garlic; sauté until tender but not brown. Add green pepper, tomato sauce, tomatoes, olives, salt, and sugar.

Cover and simmer about 25 minutes. Stir in mussels, sherry, and butter. Heat thoroughly, stirring. Serve over hot rice; sprinkle with parsley.

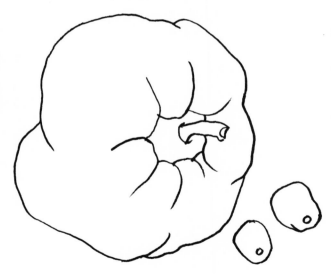

New Hangtown Fry

Serves 4

Brunch at our house would not be complete without this delicious and easy-to-fix egg dish.

Pat mussels dry with paper towel. Beat eggs with cream, Worcestershire sauce, salt, and pepper.

Fry bacon in large frying pan until crisp, drain on paper towel; pour fat out of pan.

Melt butter in same frying pan over medium heat and arrange mussels evenly in it. Lay bacon over mussels, then pour eggs in. Cook until eggs are set. If necessary, slip pan under preheated broiler to finish cooking top. Sprinkle with paprika. Serve with muffins and butter.

1 cup *(250 grams)* **cooked and drained mussel meats (canned, or fresh steamed by method on page 31)**
8 eggs
¹/₄ cup *(¹/₂ deciliter)* **cream (any kind)**
1 teaspoon Worcestershire sauce
Salt and pepper to taste
8 slices bacon
2 tablespoons butter or margarine
Paprika
Suggested accompaniment: hot corn muffins and butter

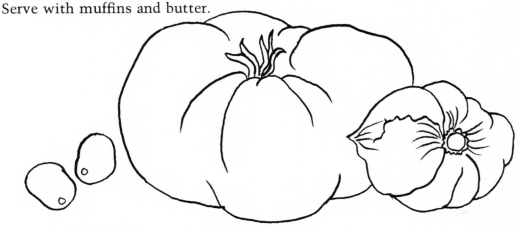

Saquish Fritters

Serves 8 [Q]

2 cups *(280 grams)* **flour sifted with 2 teaspoons baking powder**
¹/₂ cup *(1 deciliter)* **strained mussel broth saved from steaming, or canned or bottled clam juice**
¹/₄ cup *(¹/₂ deciliter)* **milk**
3 eggs, separated
2 cups *(500 grams)* **cooked, chopped, and drained mussel meats (canned, or fresh steamed by method on page 31)**
Salad oil for deep-frying
Parsley sprigs and lemon wedges

Saquish is the Indian word for clam or mussel. It is also the name given to a promontory of land extending into Plymouth Bay, Massachusetts, where Thorwald, son of Eric the Red, was buried in A.D. 1004.

Blend flour mixture with broth, milk, and egg yolks until smooth; add mussels. Beat egg whites until stiff and fold in.

Warm a serving platter in the oven. Heat oil to 375° F *(190° C)* and drop in fritters by the spoonful, a few at a time; fry until light brown, about 3 minutes. Drain on paper towels and keep warm in oven until all are cooked. Garnish platter with parsley and lemon.

Down-Home, Everyday Main Dishes

Musselburgers

Spread bun tops and bottoms with butter. Blend mussels, cream cheese, chives, and lemon juice. Spread mixture on bun bottoms and put under preheated broiler until cheese is hot (do not let brown). Toast bun tops.

On each bun bottom place tomato and 2 bacon slices; cover with tops. Serve warm, with pickles.

4 hamburger buns, each split in half
Butter or margarine
1 cup *(250 grams)* **cooked, chopped, and drained mussel meats (canned, or fresh steamed by method on page 31)**
3-ounce *(90-gram)* **package softened cream cheese**
1 tablespoon chopped chives or minced green onion
1 tablespoon lemon juice
4 slices tomato
8 slices bacon, cooked crisp
Dill pickles

Crispy Crumb Cakes

Serves 4 [Q]

1 cup *(250 grams)* **cooked, chopped, and drained mussel meats (canned, or fresh steamed by method on page 31)**
1 teaspoon lemon juice
1 teaspoon Worcestershire sauce
3 tablespoons minced celery
1 egg yolk
1½ cups *(180 grams)* **freshly made bread crumbs**
Salt and pepper to taste
1 whole egg, beaten
Fine dry bread crumbs for coating
Salad oil for deep-frying
Tomato sauce or tartar sauce
Lemon wedges

Combine mussels, lemon juice, Worcestershire sauce, celery, egg yolk, and freshly made bread crumbs. Beat until smooth. Season with salt and pepper. Shape into 8 balls or cakes.

Dip each in the whole egg, then roll in dry bread crumbs. Refrigerate an hour or more to dry and firm. Let come to room temperature before cooking. Fry a few at a time in oil heated to 375° F *(190° C)*. Drain on paper towels. Serve hot with heated tomato sauce, or at room temperature with tartar sauce. In either case, provide lemon.

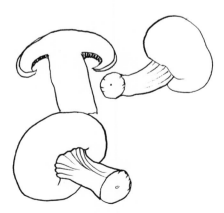

Mushroom Croquettes

In large frying pan, melt butter over low heat and blend in flour; remove from heat. Add mussel broth and cook, stirring constantly, until thickened. Season with salt and pepper. Beat a little sauce into egg yolks, then blend gradually into sauce. Remove from heat.

Add mussels, rice, mushrooms, parsley, cheese, and cayenne. Cool, then shape into 12 croquettes and roll each in crumbs.

Warm tomato sauce in saucepan. Heat oil to 375° F *(190° C)* in deep fryer. Fry a few croquettes at a time in oil until brown and crispy. Serve at once with sauce ladled over.

2 tablespoons butter or margarine
2 tablespoons flour
1 cup *(2½ deciliters)* **strained mussel broth saved from steaming, or canned or bottled clam juice**
Salt and pepper to taste
2 egg yolks, beaten
2 cups *(500 grams)* **cooked, chopped, and drained mussel meats (canned, or fresh steamed by method on page 31)**
2 cups *(300 grams)* **cooked rice**
½ pound *(250 grams)* **fresh mushrooms, chopped, or an 8-ounce *(250-gram)* can sliced mushrooms, drained**
1 tablespoon chopped parsley
½ cup *(50 grams)* **shredded Gruyère or Swiss cheese**
Cayenne
Fine dry bread crumbs for coating
About 1½ cups *(3½ deciliters)* tomato sauce
Salad oil for deep-frying

Old-Fashioned Hash

Serves 4 [Q]

1 cup *(250 grams)* **cooked, chopped, and drained mussel meats (canned, or fresh steamed by method on page 31)**
1 **medium-sized onion, chopped**
2 cups *(250 grams)* **chopped parboiled potatoes**
1 **tablespoon minced green pepper**
1 **tablespoon minced pimiento**
Strained mussel broth saved from steaming (or a little milk, cream, or water)
¼ **cup** *(65 grams)* **butter or margarine**
Salt and pepper to taste
Parsley for garnish

Combine mussels, onion, potatoes, green pepper, pimiento, and enough broth to bind mixture.

In a large frying pan, melt butter and spread hash evenly in it. Sprinkle with salt and pepper. Cook over low heat. Meanwhile, warm a serving platter in the oven. When crust has formed on bottom of hash, fold over and slide onto platter. Garnish with parsley.

Down-Home, Everyday Main Dishes

Poultry Stuffing

This makes enough stuffing for a 12-pound turkey. Those who like oysters in stuffing surely will enjoy the rich flavor of mussels instead. To this recipe you may add chopped onion, walnuts, an egg or two, or any favorite stuffing ingredient.

Combine all ingredients in bowl and toss with fork; take care not to mash or compress.

1 cup *(250 grams)* **cooked, chopped, and drained mussel meats (canned, or fresh steamed by method on page 31)**
6 cups *(720 grams)* **soft fresh bread crumbs**
2 cups *(5 deciliters)* **strained mussel broth saved from steaming, or canned or bottled clam juice**
1 cup (¹/₂ **pound, or** *250 grams*) **butter or margarine, melted**
2 cups *(240 grams)* **finely chopped celery**
¹/₄ **teaspoon ground sage**
Salt, pepper, and paprika to taste

Casseroles American Style

Probably more than any other dish, the casserole is identified with the American way of cooking. Although the basic style is American, the dishes in this section can be adapted to the flavor of many countries by varying the choice of seasonings.

Quick Casserole

Serves 4

Butter a wide, shallow baking dish and spread mussels in it. Mix onion, garlic, parsley, paprika, oregano, and cheese; sprinkle over mussels. Add salt and pepper. Top with crumbs and drizzle butter over crumbs. Bake uncovered in preheated oven at 425° F *(215° C)* for 20 minutes, or until thoroughly hot and brown on top.

3 cups *(750 grams)* **cooked and drained mussel meats (canned, or fresh steamed by method on page 31)**
1 teaspoon grated onion
1 clove garlic, puréed
1 tablespoon chopped parsley
¹/₂ teaspoon paprika
¹/₂ teaspoon crumbled oregano
¹/₄ **cup** *(25 grams)* **grated Parmesan cheese**
Salt and pepper to taste
¹/₂ **cup** *(60 grams)* **fine dry bread crumbs**
¹/₄ **cup** *(65 grams)* **butter or margarine, melted**

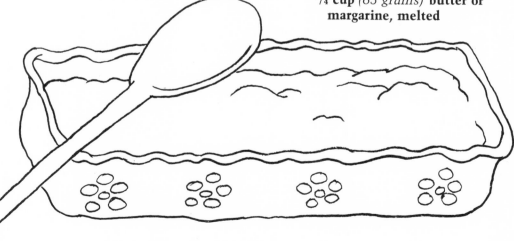

Mussels Rosé

12 hard-boiled eggs, each cut
  in half lengthwise
1 tablespoon softened butter
1 cup *(250 grams)* cooked
  mussel meats (canned, or
  fresh steamed by method on
  page 31), chopped and
  drained
Salt and pepper to taste
2 tablespoons butter or margarine
2 tablespoons flour
1 cup *(2½ deciliters)* milk
Pinch of mace or nutmeg
¼ cup *(½ deciliter)* tomato
  sauce
Fine dry bread crumbs
Parsley, watercress, or lemon
  wedges for garnish
Fried shoestring potatoes

Mash egg yolks with the 1 tablespoon butter; mix in mussels. Season with salt and pepper. Fill egg-white halves with yolk-mussel mixture and place in shallow baking pan.

Melt the 2 tablespoons butter in saucepan, remove from heat, and blend in flour. Add milk, stir, and return to heat. Cook, stirring constantly, until thick. Season with salt, pepper, and mace. Blend in tomato sauce; pour over eggs. Sprinkle with bread crumbs.

Bake uncovered in preheated oven at 350° F *(180° C)* for 20 minutes, or until thoroughly hot. Add garnish. Serve with potatoes.

Noodle Bake

Serves 8

Place cooked noodles and mussels in a 4-quart *(4-liter)* casserole.

In large frying pan, melt butter and add mushrooms and green pepper; sauté until pepper is tender. Remove from heat; with slotted spoon remove vegetables and put into casserole.

Into remaining butter, blend flour, then milk. Return to low heat, add salt, pepper to taste, and stir constantly until thick. Remove from heat, stir in wine, and pour sauce into casserole.

Toss all ingredients to coat with sauce, sprinkle cheese on top, and color cheese with paprika. Bake uncovered in pre-heated oven at 350° F *(180° C)* about 45 minutes, or until thoroughly hot and cheese is lightly browned. (*Note:* Casserole may be frozen before baking. Add 30 minutes to baking time when cooking from the frozen state.)

12 ounces *(375 grams)* **thin noodles, cooked and drained**
2 cups *(500 grams)* **cooked and drained mussel meats (canned, or fresh steamed by method on page 31)**
$^{1}/_{2}$ cup (**$^{1}/_{4}$ pound, or** *125 grams*) **butter or margarine**
1 cup *(60 grams)* **sliced fresh mushrooms or an 8-ounce** *(250-gram)* **can sliced mushrooms, drained**
2 tablespoons chopped green pepper
6 tablespoons flour
4 cups *(1 liter)* **milk**
$^{1}/_{2}$ teaspoon salt
Black pepper
$^{3}/_{4}$ cup *(2 deciliters)* **dry white wine**
2$^{1}/_{2}$ cups (**$^{1}/_{2}$ pound, or** *250 grams*) **shredded Swiss cheese**
Paprika

Cheesy Rice Casserole

Serves 12

¹/₄ cup *(65 grams)* **butter or margarine**
¹/₄ cup *(35 grams)* **flour**
1 teaspoon salt
¹/₈ teaspoon black pepper
2 cups *(5 deciliters)* **strained mussel broth saved from steaming, or canned or bottled clam juice**
2 cups *(5 deciliters)* **milk**
Large pinch of saffron (optional)
3 cups *(450 grams)* **cooked rice**
4 cups *(1 kilogram)* **cooked and drained mussel meats (canned, or fresh steamed by method on page 31)**
¹/₂ **cup** *(80 grams)* **chopped pimiento-stuffed olives**
3 tablespoons pimiento cut in strips
2¹/₂ cups (¹/₂ pound, or *250 grams***) shredded American cheese**

In saucepan over low heat, melt butter and blend in flour. Remove from heat; add salt, pepper, broth, milk, and saffron. Cook, stirring constantly, until thick.

Butter a 4-quart *(4-liter)* casserole. Arrange layers of rice, mussels, olives and pimiento, cheese, and sauce, ending with cheese. Bake uncovered in preheated oven at 350° F *(180° C)* for 30 minutes, or until browned on top.

New England Casserole with Cranberries

Serves 6

Steam mussels open in wine according to instructions on page 31. Discard shells. Strain broth and reserve 1 cup *(2½ deciliters)*. Put mussels in buttered 2-quart *(2-liter)* baking dish. Cover with cranberry sauce.

In saucepan, melt butter; add mushrooms, onion, and celery; sauté until tender. Stir in flour to coat vegetables, then add reserved broth and cream; heat, stirring constantly, until thick.

Pour vegetable sauce over mussels and cranberries. Top with mixture of crumbs and cheese. Bake uncovered in preheated oven at 375° F *(190° C)* for about 30 minutes, or until hot and browned on top.

2 quarts *(2 liters)* **uncooked mussels in their shells, cleaned according to instructions on page 25**
1 cup *(2½ deciliters)* **dry white wine**
1 cup *(2½ deciliters)* **jellied cranberry sauce**
¼ cup *(65 grams)* **butter or margarine**
½ pound *(250 grams)* **fresh mushrooms, sliced**
1 medium-sized onion, chopped
⅓ cup *(50 grams)* **chopped celery**
¼ cup *(35 grams)* **flour**
1 cup *(2½ deciliters)* **heavy (whipping) cream**
2 cups *(200 grams)* **cracker crumbs**
½ cup *(50 grams)* **shredded Gruyère or mild cheddar cheese**

California Casserole

Serves 12

½ cup (¼ **pound, or** *125 grams*) **butter or margarine**
1 **large eggplant, peeled and thinly sliced**
3 **large zucchini, peeled and thinly sliced**
4 **cups** (*1 kilogram*) **cooked and drained mussel meats (canned, or fresh steamed by method on page 31)**
2 **teaspoons lemon juice**
Salt to taste
¼ **teaspoon nutmeg**
1 **can (14 to 16 ounces, or** *400 to 500 grams*) **artichoke hearts, drained and sliced**
½ **cup** (*70 grams*) **flour**
1 **quart** (*1 liter*) **milk, scalded**
2 **cups (about 1 pound, or** *500 grams*) **ricotta cheese**
4 **eggs, beaten**

This recipe can be prepared the day before and baked prior to serving.

In large frying pan, melt a fourth of the butter over medium-high heat and sauté eggplant until tender; remove. Melt another fourth of the butter and sauté zucchini until golden. Season mussels with lemon juice, salt, and nutmeg.

Butter a deep skillet or casserole of at least 5-quart (*5-liter*) capacity. In it alternate layers of mussels, eggplant, zucchini, and artichokes. Cover and refrigerate overnight.

Just before baking, melt remaining butter in large pan over low heat and stir in flour; remove from heat. Add hot milk and cook, stirring constantly, until smooth and thickened; add salt and remove from heat. Beat cheese into sauce, then add eggs gradually; correct salt.

Drain liquid from casserole by tipping dish and spooning out excess liquid. Mask with sauce. Bake uncovered in preheated oven at 375° F (*190° C*) for about 30 minutes, until hot and brown on top.

Casseroles American Style

Vegetable Pie

Serves 6

Cook potatoes and onions in broth until almost tender. In a separate large saucepan, melt butter and blend in flour; remove from heat and add milk and cream. Return to heat and cook, stirring constantly, until thick. Stir in mussels and potatoes with broth. Add peas, basil, salt, and pepper; simmer just until peas are brilliant green.

Pour into buttered shallow baking dish and cover with crumbs. (At this point, pie may be frozen for future baking.) Bake uncovered in preheated oven at 425° F *(215° C)* for 20 minutes, or until bubbly and lightly browned. Add 15 minutes to baking time when cooking from the frozen state.

2 large potatoes, peeled and thinly sliced
2 large onions, thinly sliced
1 cup *(2½ deciliters)* strained mussel broth saved from steaming, or canned or bottled clam juice
2 tablespoons butter
2 tablespoons flour
2 cups *(5 deciliters)* milk
½ cup *(1¼ deciliters)* heavy (whipping) cream
1½ cups *(375 grams)* cooked mussel meats (canned, or fresh steamed by method on page 31), cut in small pieces
½ cup *(65 grams)* frozen green peas
½ teaspoon crumbled basil
Salt and pepper to taste
1 cup *(100 grams)* cracker crumbs coated with 2 or 3 tablespoons melted butter

Cape Cod Croquettes

Serves 6

2 cups *(500 grams)* **cooked, chopped, and drained mussel meats (canned, or fresh steamed by method on page 31)**
1 cup *(60 grams)* **finely chopped fresh mushrooms**
1 tablespoon **chopped parsley**
¹/₂ cup *(50 grams)* **shredded Gruyère or Swiss cheese**
1 egg, **beaten**
1 cup *(120 grams)* **fine dry bread crumbs**
Salad oil for deep-frying
Sprigs of parsley
Lemon wedges
Tomato seafood sauce, optional (recipe with Batter-Fried Tidbits on page 61)

Combine mussels, mushrooms, parsley, and cheese; form into croquettes. Roll in egg, then in crumbs. Place on plate to let dry and firm for an hour or more.

Deep-fry in oil heated to 375° F *(190° C)* until light brown. Serve hot on bed of parsley with lemon wedges and, if you like, tomato seafood sauce—cold or heated.

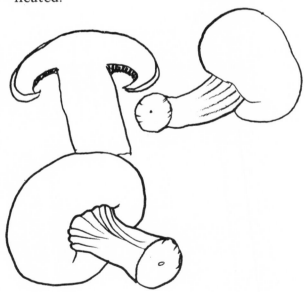

124

Main Dishes with a Foreign Accent

Why not honor your guests with an exciting main dish! This collection from various countries of the world will add flair and variety to your entertaining as well as heighten the interest in your everyday menu. You would surely find these dishes in restaurants and homes if you were to travel through their countries of origin.

Mouclade

We found this such a delectable dish at the Hôtel d'Angleterre in La Rochelle, France, that one serving was not enough.

Steam mussels, using the wine as liquid, according to instructions on page 31. Warm a serving platter in the oven. Discard one shell from each mussel and put mussels with remaining shells on platter. Cover with damp cloth or lid to keep warm. Strain mussel broth through several thicknesses of cheesecloth or paper coffee filters; reserve 1 cup *(2½ deciliters)* for this recipe.

In saucepan over medium heat, melt butter and sauté onions until golden brown; stir in flour. Add the reserved broth gradually and cook, stirring constantly, until thickened.

Remove from heat. Beat cream with egg yolks; to sauce add cream mixture, parsley, and salt and pepper to taste. Heat just enough, stirring constantly, to be very hot; do not let simmer. Spoon over warm mussels. Serve immediately.

4 quarts *(4 liters)* **uncooked mussels in their shells, cleaned and shells well scrubbed according to instructions on page 25**
½ cup *(1 deciliter)* **dry white wine**
¼ cup *(65 grams)* **butter or margarine**
2 medium-sized onions, finely chopped
2 tablespoons flour
1 cup *(2½ deciliters)* **heavy (whipping) cream**
2 egg yolks
1 tablespoon chopped parsley
Salt and pepper

Mussels à la Poulette

Serves 4

3 quarts *(3 liters)* **uncooked mussels in their shells, cleaned and shells well scrubbed according to instructions on page 25**
2 tablespoons butter or margarine
2 tablespoons flour
³/₄ cup *(2 deciliters)* all-purpose cream or half-and-half
1 egg yolk, beaten
1 teaspoon lemon juice
2 tablespoons finely chopped parsley
Salt and pepper
Onion-mushroom garnish

Small onions and mushrooms sautéed in butter are the traditional garnish for à la poulette *dishes.*

Steam mussels according to instructions on page 31. Strain broth and reserve 1 cup *(2¹/₂ deciliters)* for this recipe. Keep mussels warm in a covered pan in the oven.

In a saucepan, melt butter and stir in flour until it foams. Remove from heat; add the reserved broth and half the cream. Beat rest of cream into egg yolk.

Return pan to heat and cook, stirring constantly, until thickened. Slowly stir in cream and yolk, then lemon juice, parsley, and salt and pepper to taste. Heat, stirring, just until hot.

Divide mussels into soup plates. Spoon sauce over them and serve at once with garnish if desired.

Moules
Marinière

This French way of preparing mussels in a wine and onion sauce has become the most widely known to European and American gourmets.

Warm a tureen or individual soup plates in the oven. Into a large saucepan put wine, water, onion, celery, bay leaf, and mussels. Steam according to instructions on page 31. Put mussels still in shells in tureen or bowls; cover with damp cloth or lid to keep warm. Reserve broth.

In small saucepan over low heat, melt butter, add flour, and stir until blended; remove from heat. Strain mussel broth through several thicknesses of cheesecloth into flour mixture. Return to low heat and cook, stirring constantly, until thick. Add parsley, salt, and pepper. Pour sauce over warm mussels. Serve at once.

1 cup *(2 1/2 deciliters)* **dry white wine**
1 cup *(2 1/2 deciliters)* **water**
1 **medium-sized onion, chopped**
1 **stalk celery, sliced across**
1 **bay leaf**
4 **quarts** *(4 liters)* **uncooked mussels in their shells, cleaned and shells well scrubbed according to instructions on page 25**
3 **tablespoons butter or margarine**
2 **tablespoons flour**
3 **tablespoons chopped parsley**
Salt and pepper to taste

Soufflé with Hollandaise

Serves 4

3 tablespoons butter or margarine
3 tablespoons flour
1 cup *(2¹/₂ deciliters)* milk
²/₃ cup *(165 grams)* cooked,
chopped, and very well
drained mussel meats
(canned, or fresh steamed by
method on page 31)
1 teaspoon lemon juice
¹/₂ teaspoon dried tarragon
Dash of cayenne
Salt and pepper to taste
4 eggs, separated
Hollandaise sauce (recipe
below employs classic
method, or see recipe for
Quick Blender Hollandaise
with Baked Tomato Brunch
on page 157)

Make a white sauce by melting butter in saucepan and stirring in flour; remove from heat. Gradually blend in milk and cook, stirring constantly, until thick. Add mussels, lemon juice, tarragon, cayenne, salt, and pepper. Beat egg yolks and blend well into mussel mixture. This much may be done ahead.

When ready to bake, beat egg whites with a dash of salt until they hold stiff peaks but do not look dry. Gently fold into sauce. Pour into buttered 1¹/₂-quart *(1¹/₂-liter)* soufflé dish.

Bake in preheated oven at 350° F *(180° C)* for about 35 minutes, until puffy and golden brown. Center of soufflé should feel nonliquid but not too firm. Serve immediately with warm hollandaise sauce.

Hollandaise Sauce

For best results ingredients should be cold. Put half the butter in small heavy saucepan over low heat. Add lemon juice and egg yolks. Cook, stirring rapidly with wire whisk, until butter melts. Add rest of butter and continue whisking until sauce thickens.

Remove from heat and stir in salt and cayenne. Serve at once or keep warm over hot (not boiling) water. If sauce separates, blend in a few drops of hot water while whisking. Makes about ¹/₂ cup *(1 deciliter).*

¹/₂ **cup (¹/₄ pound, or** *125 grams*) **butter or margarine**
1 tablespoon lemon juice
2 egg yolks, beaten
¹/₂ **teaspoon salt**
Dash of cayenne

Main Dishes with a Foreign Accent

Mussels Genève

2 quarts *(2 liters)* **uncooked mussels in their shells, cleaned and shells well scrubbed according to instructions on page 25**
1¹/₂ cups *(3¹/₂ deciliters)* **beer**
2 tablespoons butter or margarine
1 cup *(2¹/₂ deciliters)* **yogurt**
¹/₂ cup *(1 deciliter)* **heavy (whipping) cream**
1 teaspoon meat extract or 1 bouillon cube
2 teaspoons Worcestershire sauce
1 tablespoon lemon juice
1 teaspoon dried tarragon
2 tablespoons finely chopped parsley
Salt and pepper to taste
Paprika
Suggested accompaniment: boiled potatoes and French-cut green beans

Put mussels, beer, and butter in large saucepan. Cover and steam according to instructions on page 31. Discard one shell from each mussel; put remaining shells with mussels in shallow baking dish.

Combine yogurt, cream, meat extract, Worcestershire sauce, lemon juice, tarragon, and parsley. Season with salt and pepper. Pour over mussels, sprinkle with paprika.

Bake uncovered in preheated oven at 425° F *(215° C)* for 10 to 15 minutes, or until hot and bubbly. Serve with vegetables.

Mussels à la Brussels

Serves 4 to 6

In saucepan over low heat, melt butter and blend in flour. Remove from heat; add parsley, onion, Worcestershire sauce, and milk. Cook, stirring constantly, until thick. Season with salt and pepper; add mussels.

Pour into buttered shallow baking dish and cover with crumbs. Bake uncovered in preheated oven at 375° F *(190° C)* for 20 minutes, or until crumbs are brown.

1/4 **cup** *(65 grams)* **butter or margarine**
1/4 **cup** *(35 grams)* **flour**
1 **tablespoon chopped parsley**
2 **teaspoons minced onion**
1 **tablespoon Worcestershire sauce**
2 **cups** *(5 deciliters)* **milk**
Salt and pepper to taste
2 **cups** *(500 grams)* **cooked and drained mussel meats (canned, or fresh steamed by method on page 31)**
1 **cup** *(120 grams)* **bread crumbs (fine dry or freshly made) blended with 2 or 3 tablespoons melted butter**

Crêpes Stuffed with Crab

Serves 4

¹/₂ **cup** *(70 grams)* **flour**
¹/₄ **teaspoon salt**
2 eggs, beaten
⁷/₈ **cup** *(2¹/₄ deciliters)* **milk**
¹/₂ **tablespoon butter or margarine, melted**

Combine flour, salt, eggs, milk, and butter. Stir until batter is thin and creamy. Refrigerate 2 hours.

Use small frying pan, griddle, or crêpe pan. Lightly grease before each pancake with butter or oil on crumpled paper towel. Use ¹/₄ cup *(¹/₂ deciliter)* batter for each pancake. Cook on medium heat. When bubbles form on surface, turn and cook other side until golden brown. (*Note:* To speed cooking if crêpe pan is used, have large frying pan also heated—turn pancake into frying pan to cook second side.)

Wrap the 4 crêpes in foil and keep warm in oven. Or freeze and use as needed.

Filling

Make sauce by melting butter in sauce-pan, then stirring in flour until it foams. Remove from heat, gradually blend in milk. Cook, stirring constantly, until thick. Stir in sherry and cheese until cheese melts. Remove from heat and ladle a third of the sauce into bowl; reserve.

To sauce remaining, add mussels, crab, onion, lemon juice, salt, and pepper.

Butter shallow baking dish. Spoon generous portions of mussel mixture on each crêpe, roll up, and arrange side by side (seam side down) in dish. Pour reserved sauce over all and sprinkle with paprika.

Bake uncovered in preheated oven at 375° F *(190° C)* for about 20 minutes, or until hot and bubbly.

4 tablespoons butter or margarine
6 tablespoons flour
2 cups *(5 deciliters)* **milk**
¹/₃ cup *(³/₄ deciliter)* **dry sherry**
³/₄ cup *(75 grams)* **shredded Gruyère or Swiss cheese**
1 cup *(250 grams)* **cooked, finely chopped, and drained mussel meats (canned, or fresh steamed by method on page 31)**
1 cup *(250 grams)* **finely chopped cooked crab**
1 tablespoon grated onion
1 tablespoon lemon juice
Salt and pepper to taste
Paprika

Individual Mussel Quiches

Serves 6

1 cup (¹/₂ **pound,** or *250 grams*) **butter or margarine, softened**
8-ounce *(250-gram)* **package cream cheese, at room temperature**
¹/₄ **cup** *(¹/₂ deciliter)* **heavy (whipping) cream**
2¹/₂ **cups** *(350 grams)* **sifted flour**
1 **teaspoon salt**

Blend butter, cheese, and cream. Sift flour and salt into cheese mixture; blend thoroughly. Divide dough and press with fingers to line 12 individual tart pans.

¹/₂ **cup** *(125 grams)* **cooked, chopped, and drained mussel meats (canned, or fresh steamed by method on page 31)**
1¹/₂ **cups** *(3¹/₂ deciliters)* **heavy (whipping) cream**
3 **eggs, beaten**
¹/₂ **cup** *(50 grams)* **shredded Gruyère or Swiss cheese**
1 **teaspoon lemon juice**
1 **teaspoon dried dill weed**
Pinch of cayenne
Grated Parmesan cheese

Filling

Be sure mussels are thoroughly drained, pat dry, and divide into pastry-lined pans. Mix cream, eggs, cheese, lemon juice, dill weed, and cayenne. Spoon over mussels, filling pans no more than two-thirds full. Sprinkle with Parmesan cheese.

Bake for 5 minutes in preheated oven at 425° F *(215° C)*, then reduce temperature to 350° F *(180° C)* and bake about 15 minutes more, or until filling has firmed. (Bake only 10 minutes if you want to freeze quiches and reheat later.)

Main Dishes with a Foreign Accent

Moules à la Hongroise

Serves 4

Steam mussels according to instructions on page 31, first placing wine, water, onion, celery, and bay leaf in the pan, then mussels and 1 tablespoon of the paprika.

Remove and discard one shell from each mussel. Put mussels in half shells in shallow baking dish suitable for taking to the table; cover and keep warm in oven. Strain broth and measure 1 cup (2½ *deciliters*) for this recipe.

Melt butter in saucepan; add onion and garlic, sauté until tender. Add remaining 1 tablespoon paprika. Remove from heat.

Dissolve flour in cream. Gradually stir cream and reserved broth into seasoned onions. Return to heat and cook, stirring constantly, until thick. Add salt and pepper.

Pour sauce over mussels. Sprinkle on parsley. Serve piping hot.

1 **cup** *(2½ deciliters)* **dry white wine**
1 **cup** *(2½ deciliters)* **water**
1 **medium-sized onion, thinly sliced or chopped**
1 **stalk celery, sliced across**
1 **bay leaf**
4 **quarts** *(4 liters)* **uncooked mussels in their shells, cleaned and shells well scrubbed according to instructions on page 25**
2 **tablespoons paprika**
3 **tablespoons butter or margarine**
½ **cup** *(70 grams)* **finely chopped onion**
1 **clove garlic, minced or puréed**
3 **tablespoons flour**
½ **cup** *(1 deciliter)* **light cream or half-and-half**
Salt and pepper to taste
2 **tablespoons chopped parsley**

Midia Dolma

24 cooked and drained mussels in their shells (canned, or fresh steamed by method on page 31)

3 cups *(7½ deciliters)* strained mussel broth saved from steaming, or canned or bottled clam juice

⅓ cup *(1 deciliter)* olive oil

½ cup *(120 grams)* uncooked rice

1 medium-sized onion, chopped

½ teaspoon allspice

½ teaspoon cinnamon

¼ teaspoon freshly ground black pepper

¼ cup *(50 grams)* pine nuts (or walnuts may be substituted)

¼ cup *(50 grams)* dried currants

Lemon wedges for garnish

In Greece, the small mussels called kydonia *are sold in villages along the edge of the sea.* Dolmas *are fig or grape leaves rolled and stuffed with bits of meat, rice, currants, nutmeats, and spices. In this recipe mussel shells take the place of leaves and are stuffed as the Byzantines might have done.*

Remove cooked mussels in shells from their broth and set aside. Strain broth through coffee filters or several layers of cheesecloth; reserve.

Heat oil in medium saucepan, and cook rice and onion in it until golden brown. Add 1 cup *(2½ deciliters)* mussel broth, seasonings, nuts, and currants. Bring to a boil over high heat, then cover and lower heat. Simmer for 15 minutes or until all liquid is absorbed. Remove stuffing from heat.

Remove top shell from each mussel and set aside. Cover each remaining mussel in its half shell with stuffing. Replace top shell and press down. Place in shallow casserole with 2 cups *(5 deci-*

liters) of broth; add water if there is not enough broth. Cover and bake in pre-heated oven at 375° F *(190° C)* for 45 minutes until water is absorbed. Refrigerate when cool.

Serve cold, garnished with lemon wedges.

Main Dishes with a Foreign Accent

Dutch Kebabs

**2 tablespoons butter or marga-
  rine**
1 tablespoon minced onion
1/2 tablespoon minced garlic
1/2 cup *(1 deciliter)* **water**
2 tablespoons white vinegar
4 tablespoons peanut butter
Cayenne or crushed red chili
Salt and sugar to taste
**40 cooked whole mussel
  meats (canned, or fresh
  steamed by method on page
  31)**
1 cup *(120 grams)* **fine dry
  bread crumbs**
Salad oil for deep-frying
**Paprika and parsley sprigs for
  garnish**

*This recipe, resembling a Javanese sate,
is typical of some dishes of Indonesian
origin that have been adopted by the Dutch.
You will need 8 wooden or bamboo skewers.*

For sauce, melt butter in saucepan and
sauté onion and garlic in it until onion is
tender and golden. Add water, vinegar,
peanut butter, and cayenne or chili to
taste (some people like this type dish
very hot). Cook, stirring often, until
sauce is thick. Add salt and a little sugar
if desired. Keep warm.

Arrange 5 mussels on each skewer; roll
in crumbs. Fry in oil heated to 375° F
(190° C) until crumbs are brown. Meanwhile, warm a serving platter.

To serve, arrange skewers on platter
side by side. Spoon sauce on mussels (not
too much—just enough to enhance, not
mask completely). Sprinkle with paprika;
add parsley.

Swedish Pudding

Serves 4

This recipe is a variation of a dish served on Swedish steamship lines.

Butter a small shallow baking dish, about 1½-quart *(1½-liter)* capacity. Arrange potatoes and mussels in layers ending with potatoes. Blend cream, eggs, melted butter, and dill seed. Pour over potatoes. Sprinkle with paprika.

Bake uncovered in preheated oven at 325° F *(165° C)* for 45 minutes, or until pudding has consistency of custard.

4 cups *(500 grams)* **boiled potatoes, sliced**
1 cup *(250 grams)* **cooked and drained mussel meats (canned, or fresh steamed by method on page 31)**
2 cups *(5 deciliters)* **light cream or half-and-half**
2 eggs, beaten
2 tablespoons butter or margarine, melted
1 teaspoon dill seed
Paprika

Paella

3 quarts *(3 liters)* **cooked and drained mussels in their shells (canned, or fresh steamed by method on page 31)**

About 4¹/₂ cups *(11 deciliters)* **strained mussel broth saved from steaming (add water if not enough broth)**

¹/₄ **cup** *(¹/₂ deciliter)* **olive oil**

¹/₂ **cup** *(70 grams)* **minced onion**

2 large cloves garlic, crushed

3 cups *(720 grams)* **uncooked rice**

1 teaspoon saffron

Salt to taste

1 cup *(125 grams)* **uncooked green peas, fresh shelled or frozen**

12 thin slices hard Spanish or Portuguese sausage

36 thin strips canned pimiento

In Spain, where mussels are intensively cultivated, they are eaten in many ways, including the dish called paella.

If you do not have a traditional paella pan, use a large shallow casserole (preferably with lid) that can be placed both on direct heat and in the oven. The traditional paella pan does not have a lid and paella is made entirely over direct heat, as are most foreign dishes that originated in pre-home-oven days.

Remove and discard one shell from 36 of the mussels; place the other 36 half shells with mussels in them in saucepan with about ¹/₂ cup *(1 deciliter)* broth or water; cover with lid and keep warm. Remove remaining mussels from their shells and discard shells.

Pour oil into the large casserole over medium heat; add onion and garlic, sauté until golden. Remove garlic and add rice. Cook, stirring, until rice is golden brown.

Meanwhile heat 4 cups *(1 liter)* broth with saffron until it turns yellow; stir into rice. Adjust seasoning with salt. Add peas, sausage, and mussel meats without shells (fold in with fork or wooden spoon as you would toss a salad, lightly, to avoid mashing ingredients).

Cover (with lid, or tightly with foil) and bake in preheated oven at 350° F *(180° C)* for about 15 minutes, or until rice absorbs all the broth. The rice should be moist but not soupy.

Serve from pan; decorate by imbedding reserved warm mussels on half shells in ring formation on the rice. Lay a pimiento strip on each mussel in half shell for garnish.

Mussel Sauce
for Spaghetti

Serves 4

Red Sauce

¹/₂ **cup** *(1 deciliter)* **olive oil**
2 small cloves garlic, minced
1 tablespoon chopped parsley
1 cup *(250 grams)* **chopped canned tomatoes**
¹/₂ **cup** *(1 deciliter)* **strained mussel broth saved from steaming, or canned or bottled clam juice**
1 cup *(250 grams)* **cooked, chopped, and drained mussel meats (canned, or fresh steamed by method on page 31)**
Salt and pepper to taste
Spaghetti and grated Parmesan or Romano cheese

"Red" and "white" clam sauces (one with tomato, the other without) are well-known Italian spaghetti toppings. Mussels provide a delicious variation.

Heat oil in saucepan; add garlic and parsley, sauté until garlic is tender. Pour in tomatoes and broth. Cover and simmer until thickened, about 25 minutes. Add mussels; season with salt and pepper. Spoon over hot spaghetti. Sprinkle with cheese.

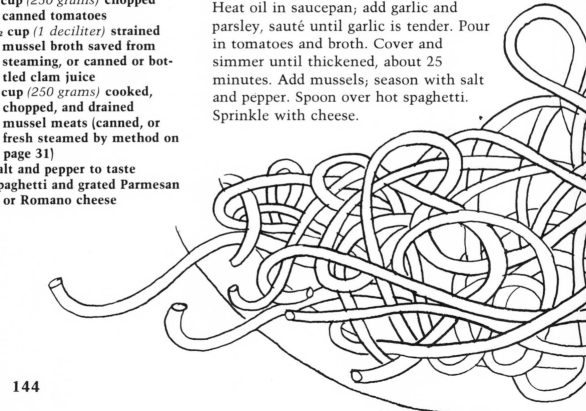

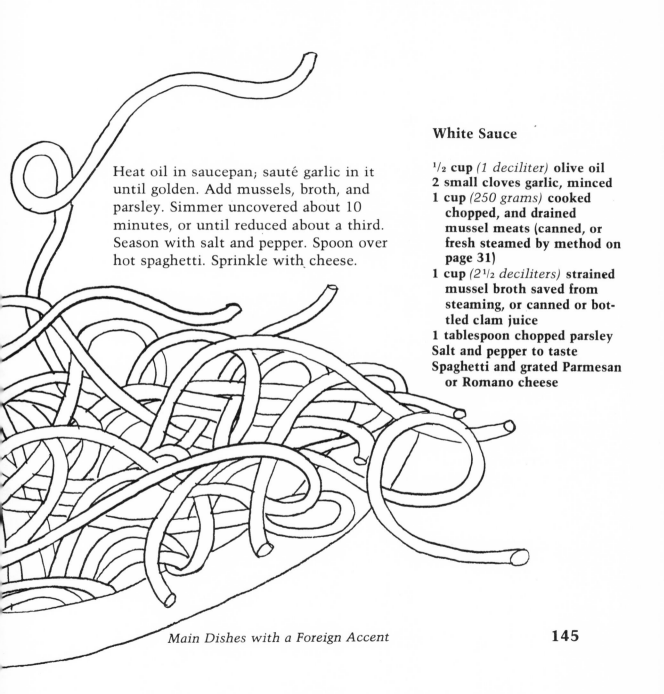

Heat oil in saucepan; sauté garlic in it until golden. Add mussels, broth, and parsley. Simmer uncovered about 10 minutes, or until reduced about a third. Season with salt and pepper. Spoon over hot spaghetti. Sprinkle with cheese.

White Sauce

¹/₂ **cup** *(1 deciliter)* **olive oil**
2 small cloves garlic, minced
1 cup *(250 grams)* **cooked chopped, and drained mussel meats (canned, or fresh steamed by method on page 31)**
1 cup *(2¹/₂ deciliters)* **strained mussel broth saved from steaming, or canned or bottled clam juice**
1 tablespoon chopped parsley
Salt and pepper to taste
Spaghetti and grated Parmesan or Romano cheese

Main Dishes with a Foreign Accent

Cannelloni Ricotta

Serves 6

1 cup *(250 grams)* cooked,
chopped, and drained
mussel meats (canned, or
fresh steamed by method on
page 31)
2 cups *(450 grams)* ricotta
cheese
¹/₃ cup *(30 grams)* chopped
parsley
2 tablespoons chopped chives
or minced green onion
2 eggs, beaten
¹/₄ teaspoon nutmeg
Salt and pepper to taste
12 cannelloni shells, cooked
according to package directions
4 cups *(1 liter)* tomato sauce
or bottled spaghetti sauce
(plain marinara style)
²/₃ cup *(70 grams)* grated
Romano cheese

Make stuffing by mixing mussels, ricotta
cheese, parsley, chives, eggs, nutmeg,
salt, and pepper. Using a small spoon or
pastry bag, stuff cannelloni shells. Butter
a shallow baking dish and lay shells side
by side. Pour on sauce and sprinkle with
cheese.

Bake uncovered in preheated oven at
400° F *(200° C)* for about 25 minutes, or
until very hot and bubbly.

Sauté Scampi Style

Serves 4

The garlic-seasoned manner in which scampi are prepared in Italy suits mussels just as well. A green salad and crusty French or sourdough bread are all you need to make a meal of this dish.

Drain raw mussels on paper toweling; coat with flour.

In large frying pan that can be taken to the table, heat oil over high heat and add butter, then garlic. Shake excess flour off mussels and slide them into oil. Sauté about 2 minutes, or until golden. Sprinkle with shallots and parsley; sauté another 2 minutes, or until liquid evaporates.

Season with plenty of pepper and salt if needed. Serve with rice and lemon.

2 quarts *(2 liters)* **uncooked mussels in their shells, cleaned and opened raw by cutting method on page 28 (discard shells)**
¼ **cup** *(35 grams)* **flour**
3 **tablespoons olive oil**
¼ **cup** *(65 grams)* **butter or margarine**
6 **cloves garlic, puréed**
2 **tablespoons minced shallots**
2 **heaping tablespoons chopped parsley**
Freshly ground pepper and salt to taste
Hot cooked rice
Lemon wedges

Thai Curry

3 tablespoons butter or margarine
1 cup *(150 grams)* chopped
 onion
1½ tablespoons curry powder
2 cups *(500 grams)* cooked
 and drained mussel meats
 (canned, or fresh steamed by
 method on page 31)
2 cups *(5 deciliters)* strained
 mussel broth saved from
 steaming, or canned or bottled clam juice
½ pound *(250 grams)* fresh
 mushrooms, sliced
4 tablespoons flour
½ cup *(1 deciliter)* cream (any
 kind)
Salt to taste
Hot cooked rice
Condiments: chopped hard-
 boiled eggs, cashews,
 raisins, bananas, chopped
 raw carrot, chutney, and
 shredded coconut

Melt butter in frying pan, add onion and curry powder. Sauté until onion is tender. Add mussels, broth, and mushrooms. Cover and simmer about 5 minutes.

Blend flour and cream. Pour into mussel mixture and cook, stirring constantly, until thick. Season with salt.

Serve over rice. Condiments may be offered from individual bowls or spooned at side of rice.

Chinese Stir-Fry

Serves 4 to 6

In a bowl, blend sherry, soy sauce, cornstarch, sugar, and vinegar until cornstarch is dissolved. Pour in chicken broth.

Heat oil over high heat in wok or large frying pan. Add garlic and stir-fry 1 minute. Toss in mussels and stir-fry until very hot. Do not brown.

Stir broth mixture and pour into wok with mussels. Cook, stirring constantly, until sauce is translucent. Remove garlic. Season with salt if necessary and pepper. Serve in bowl with parsley sprinkled on top. Provide hot, fluffy rice in bowl on the side.

3 tablespoons dry sherry
3 tablespoons soy sauce
3 tablespoons cornstarch
2 tablespoons sugar
2 tablespoons white vinegar
1 cup *(2½ deciliters)* **chicken broth**
3 tablespoons peanut or other vegetable oil (except olive)
1 large clove garlic, crushed
2 cups *(500 grams)* **cooked and drained mussel meats (canned, or fresh steamed by method on page 31)**
Salt and black or crushed red pepper
3 tablespoons chopped regular parsley or Chinese parsley
Hot cooked rice

Sweet-Sour Stir-Fry

Serves 4

¹/₂ **cup** *(70 grams)* **flour**
Salt and cayenne
¹/₃ **cup** *(1 deciliter)* **cold water**
Salad oil for deep-frying
¹/₄ **cup** *(40 grams)* **brown sugar**
¹/₄ **cup** *(¹/₂ deciliter)* **white vinegar**
1 **cup** *(2¹/₂ deciliters)* **strained mussel broth saved from steaming, or canned or bottled clam juice**
1 **tablespoon cornstarch**
¹/₂ **cucumber, peeled, seeded, and diced**
1 **cup** *(120 grams)* **coarsely diced celery**
1 **sweet red (bell) pepper, diced; use green pepper if red is unavailable**
1 **egg white**
3 **cups** *(750 grams)* **cooked and drained whole mussel meats (canned, or fresh steamed by method on page 31)**
1 **can (5 ounces, or** *150 grams)* **chow mein noodles**
Hot cooked rice

Make batter in bowl by seasoning flour with about ¹/₄ teaspoon salt and cayenne to taste. Beat in cold water and 2 tablespoons oil. Set aside, as long as several hours if you like.

In small bowl, combine brown sugar, vinegar, broth, and cornstarch; stir until dissolved.

Heat 2 tablespoons oil in saucepan over very high heat. Add cucumber, celery, and pepper. Stir-fry just until barely cooked, still tender-crisp. Stir broth mixture, pour in. Cook, stirring often, until thick and translucent. Keep sauce warm.

Finish batter by beating egg white stiff and folding gently into flour mixture. Warm a serving platter in the oven. Heat oil to 375° F *(190° C)* in frying pan or wok. Pat mussels dry with paper towel. Dip 3 or 4 at a time in batter and fry until light brown. Drain on paper towel placed on serving platter in warm oven.

When all mussels are cooked, heap on chow mein noodles. Pour sauce over, and serve immediately with bowls of rice.

Elegant Party Entrées

While most recipes using mussels are cooked just prior to serving, some dishes in this section can be prepared ahead and heated up in the last few moments before serving. Your party dinner can be a joy to prepare, and your menu a conversation piece.

Mussel Newburg

Serves 4

In saucepan over medium heat, melt butter and add mussels. Sauté about 5 minutes. Add wine and cook about 2 minutes more.

Add milk mixture and cook, stirring constantly, until thickened. Remove from heat. Spoon a little sauce into egg yolks, blend, and slowly beat all into sauce. Add salt, pepper, and paprika to taste. Return to heat and cook, stirring constantly, just until hot; do not overcook or egg will separate. Cover and keep warm while toasting bread. Cut toast diagonally twice to make toast points. Serve Newburg with rim of toast.

2 tablespoons butter or margarine
2 cups *(500 grams)* cooked and drained mussel meats (canned, or fresh steamed by method on page 31)
3 tablespoons dry sherry or Marsala
3 tablespoons cornstarch dissolved in 1 quart *(1 liter)* milk
2 egg yolks, beaten
Salt, pepper, and paprika
Bread for toasting

Baked Stuffed Mussels

Serves 6

2 quarts *(2 liters)* **uncooked mussels in their shells, cleaned and shells well scrubbed according to instructions on page 25**

¹/₂ **cup** *(1 deciliter)* **dry white wine**

¹/₄ **cup** *(65 grams)* **butter or margarine**

3 **tablespoons minced onion**

3 **tablespoons minced celery**

1 **tablespoon minced green pepper**

3 **tablespoons chopped parsley**

1 **cup** *(100 grams)* **cracker crumbs**

Several drops hot-pepper sauce

Salt

Butter

Lemon wedges dusted with paprika

Steam mussels, using the wine, according to method on page 31. Remove meat and reserve shells. Strain broth through several thicknesses of cheesecloth or paper coffee filters; reserve. Chop mussel meats.

In frying pan over medium heat, melt butter. Add onion, celery, and green pepper; sauté until tender. Remove from heat. Add mussels, parsley, crumbs, hot-pepper sauce, and enough reserved broth to bind the mixture. Salt to taste.

Fill all shells with mixture and arrange on baking sheets. Dot with butter. Bake in preheated oven at 450° F *(230° C)* for about 15 minutes, or until hot and light brown on top. Garnish servings with paprika-dusted lemon.

Elegant Party Entrées

Main-Dish Risotto

Serves 6

Chop tomatoes and drain; reserve. Measure ½ cup *(1 deciliter)* juice for this recipe and set aside.

Melt butter in large frying pan; add onion and garlic, sauté until tender. Add rice and stir over medium heat until moisture evaporates and grains look dry.

Add salt, chopped tomatoes, the reserved tomato juice, and ½ cup *(1 deciliter)* of the broth. Cover and simmer over low heat until liquid is absorbed, about 10 minutes. Add remaining 1 cup *(2½ deciliters)* broth, cover, and simmer until absorbed, about 15 minutes more.

With fork, lightly fold in mussels, shrimp, and cream. Cover and cook about 2 minutes more, or just until hot. Spoon into serving dish. Sprinkle with cheese and parsley.

1 can (about 1 pound, or *500 grams*) **tomatoes, preferably Italian plum type, with juices**

3 tablespoons **butter or margarine**

3 tablespoons **minced onion**

1 clove **garlic, minced**

1 cup *(240 grams)* **uncooked rice**

1 teaspoon **salt**

1½ cups *(3½ deciliters)* **strained mussel broth saved from steaming, or canned or bottled clam juice**

3 cups *(750 grams)* **cooked and drained mussel meats (canned, or fresh steamed by method on page 31)**

8 ounces *(250 grams)* **small peeled cooked shrimp**

½ cup *(1 deciliter)* **heavy (whipping) cream**

1 cup *(100 grams)* **grated Parmesan cheese**

3 tablespoons **minced parsley**

Baked Tomato Brunch

Serves 6

6 very large, ripe but firm tomatoes
Salt and pepper to taste
2 cups *(500 grams)* **small cooked and drained mussel meats (canned, or fresh steamed by method on page 31)**
2 tablespoons minced onion
6 tablespoons butter
6 eggs at room temperature
¹/₂ cup *(1 deciliter)* **hollandaise sauce, canned or made by quick blender method below**
Crumbled crisp bacon or parsley sprigs for garnish

Cut tops off tomatoes; carefully scoop out and discard pulp. Sprinkle salt and pepper inside. Stuff each with mixture of mussels, onion, and 1 tablespoon of the butter.

Arrange apart in a well-buttered baking dish and bake in preheated oven at 350° F *(180° C)* until tender but not losing shape. Remove and break an egg into each. Bake again just until eggs firm, about 10 minutes. Meanwhile, heat sauce if you like, spoon onto each tomato just before serving, and garnish with bacon or parsley.

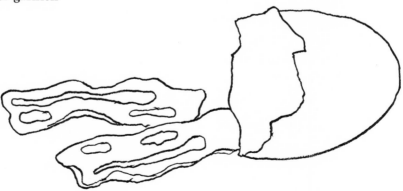

Quick Blender Hollandaise

Melt butter until bubbly, but not beginning to brown. In blender container, combine egg yolks with lemon juice. Have everything else ready and at hand. Add hot water, turn blender on high speed, and immediately begin to incorporate hot butter in a small, steady stream. When all has been poured, add salt, mustard, and pepper. Blend until well mixed and thick. Makes about 1 cup *(2½ deciliters)*. Use half for this recipe.

³/₄ cup *(190 grams)* **butter or margarine**
3 **egg yolks at room temperature**
1½ **tablespoons lemon juice**
1 **tablespoon very hot water**
½ **teaspoon salt**
About 1 **teaspoon Dijon-style mustard**
Cayenne or white pepper to taste

Elegant Party Entrées **157**

Mussels Flambées

Bread for toasting
Butter
2 tablespoons olive oil
2 cups *(500 grams)* **cooked
and drained mussel meats
(canned, or fresh steamed by
method on page 31)**
1 large green pepper, cut in
strips
1 medium-sized onion, thinly
sliced
1 jar (4 ounces, or *125 grams*)
pimiento, cut in strips and
drained
1 teaspoon Worcestershire
sauce
$^1/_4$ cup *($^1/_2$ deciliter)* brandy,
warmed in saucepan
$^1/_2$ cup *(1 deciliter)* heavy
(whipping) cream
Salt and pepper to taste

Just before preparing dish to serve, toast bread, butter it, cut each slice diagonally twice to form toast points; keep warm in oven.

At the table, heat oil in chafing dish or electric frying pan over high heat. Add mussels, green pepper, onion, pimiento, and Worcestershire sauce. Sauté until vegetables are tender but not brown, about 5 minutes. Pour warmed brandy over ingredients in pan and ignite. After flames go out, add cream, salt, and pepper. Stir until hot. Serve with border of buttered toast points.

Mussels with Green Noodles

Serves 4

In frying pan, heat olive oil and sauté onion and garlic until tender. Add tomatoes, broth, parsley, and sugar. Cook uncovered over high heat, stirring often, until slightly thickened. Add mussels; season with salt and pepper. Cover and keep warm in the oven. At the same time warm a serving dish.

Cook noodles according to package directions. Drain and toss with the sauce. Pour into serving dish, garnish with mussels in shells if available, and sprinkle plenty of cheese on top.

¹/₄ **cup** *(¹/₂ deciliter)* **olive oil**
3 tablespoons minced onion
2 cloves garlic, minced
1 can (about 1 pound, or *500 grams***) tomatoes, preferably Italian plum type, chopped and drained**
¹/₄ **cup** *(¹/₂ deciliter)* **strained mussel broth saved from steaming, or canned or bottled clam juice, or tomato juice**
1 tablespoon chopped parsley
1 teaspoon sugar
1 cup *(250 grams)* **cooked and drained mussel meats (canned, or fresh steamed by method on page 31); if fresh are prepared, save a few extra in shells for garnish**
Salt and pepper
1 pound *(500 grams)* **uncooked green (spinach) noodles**
Grated Parmesan cheese

Brochettes with Curried Rice

1½ **cups** *(360 grams)* **un-cooked rice**
1 **teaspoon curry powder**
4 **tablespoons butter or margarine, melted**
2 **cups** *(500 grams)* **cooked and drained mussel meats (canned, or fresh steamed by method on page 31)**
2 **green peppers, cut in strips or squares**
2 **red (bell) peppers, cut in strips or squares**
6 **slices raw bacon, each cut in 6 pieces**
12 **large fresh mushroom caps**
Vegetable oil
Salt and pepper to taste
Fine dry bread crumbs
Juice of 2 lemons

You will need a dozen 8-inch (20-centimeter) *skewers. This dish is a good choice in the fall when sweet red peppers come on the market. Or, at other times, you can substitute cherry tomatoes.*

Cook rice according to instructions on package, but include curry powder and 1 tablespoon of the melted butter. Keep warm.

Thread mussels on skewers alternately with green and red peppers and bacon pieces. Secure each skewer with mushroom cap. Dip each in oil, salt and pepper it, then roll in crumbs.

Broil over charcoal fire of moderate heat, brushing frequently with oil. (Or broil in oven 4 to 5 inches—*10 to 12 centimeters*—below preheated broiler.)

Serve brochettes on bed of rice which has been moistened with remaining butter mixed with lemon juice.

Sherried Shells

Serves 6

Marinate mussels in sherry for about 1 hour. Put bread in large bowl; pour in butter, cream, lemon juice, and Worcestershire sauce. (You may also add other flavorings such as a fish seasoning, herbs, or a few dashes hot-pepper sauce.)

Add mayonnaise and mix well, adding salt and pepper as you work. Fold in mussels with sherry.

Spoon mixture into large scallop shells used for baking or into individual casseroles (or even bake all in one shallow baking dish). Sprinkle on cheese and crumbs. Bake in preheated oven at 350° F *(180° C)* for 20 to 30 minutes, or until bubbly hot.

2 **cups** *(500 grams)* **cooked, chopped, and drained mussel meats (canned, or fresh steamed by method on page 31)**
$\frac{1}{2}$ **cup** *(1 deciliter)* **sherry**
3 **slices white bread, crusts removed**
$\frac{1}{4}$ **cup** *(65 grams)* **butter or margarine, melted**
$\frac{1}{2}$ **cup** *(1 deciliter)* **light cream or half-and-half**
2 **tablespoons lemon juice**
1 **teaspoon Worcestershire sauce**
$\frac{1}{2}$ **cup** *(1 deciliter)* **mayonnaise**
Salt (about $\frac{1}{2}$ teaspoon) and pepper to taste
$\frac{1}{2}$ **cup** *(50 grams)* **shredded sharp cheddar cheese**
1 **cup** *(120 grams)* **fine dry bread crumbs**

Stuffed Eggplant

Serves 6

¹/₄ **cup** *(¹/₂ deciliter)* **olive oil**
1 **clove garlic, puréed**
3 **medium-sized eggplants,**
each halved lengthwise
2 **cups** *(300 grams)* **cooked**
rice
2 **cups** *(500 grams)* **cooked**
and drained mussel meats
(canned, or fresh steamed by
method on page 31)
2 **tablespoons lemon juice**
Salt and pepper to taste
¹/₈ **teaspoon cayenne**
Dash of nutmeg
¹/₄ **cup** *(65 grams)* **butter or**
margarine
1 **egg, beaten**
Fine dry bread crumbs
3 **cups** *(7¹/₂ deciliters)* **tomato**
sauce, well seasoned with
such Mediterranean ingre-
dients as garlic, oregano,
basil, and olive oil (bottled
spaghetti sauce or Italian
cooking sauce may be used
instead)

If you can find very small eggplants and use one per serving, your dish will look even more attractive.

Put part of oil and garlic in large frying pan over medium heat; add as many eggplants as will fit, meat side down. Cook until soft enough to scoop out pulp without damaging skin; reserve. (A little browning of skin also gives good flavor to any stuffed eggplant dish.) Repeat, adding oil and garlic, until all are prepared. Finely chop pulp. Butter baking sheet; place skins side by side, open side up.

Make stuffing by mixing eggplant pulp, rice, mussels, lemon juice, salt and pepper, cayenne, and nutmeg. Heat butter in the frying pan, add stuffing, and cook until eggplant is tender, stirring frequently. Remove from heat, blend in egg, and fill eggplant skins. Sprinkle with crumbs.

Bake uncovered in preheated oven at 425° F *(215° C)* for about 35 minutes, or until very hot. Meanwhile, heat tomato sauce and warm a platter. Serve eggplants on platter with border of sauce.

Artichokes with Gruyère Sauce

Serves 6

In medium saucepan, melt the ¼ cup *(65 grams)* butter, add flour, and stir until bubbly. Remove from heat, gradually stir in milk. Cook again, stirring constantly, until thickened. Season to taste with salt and pepper. Remove from heat; stir in cheese and vermouth. Cover and keep sauce warm.

In frying pan, melt 2 tablespoons of the remaining butter and sauté artichokes until light brown. Arrange side by side in buttered shallow baking dish. Melt remaining 2 tablespoons butter in same pan over high heat and sauté mussels for 1 minute. Spoon on top of artichokes.

Spread sauce over artichokes and mussels. Bake in preheated oven at 425° F *(215° C)* for 15 to 20 minutes, or until hot and bubbly. Garnish with lemon and parsley. Serve piping hot.

¼ **cup** *(65 grams)* **butter or margarine**
¼ **cup** *(35 grams)* **flour**
2 **cups** *(5 deciliters)* **milk**
Salt and pepper
½ **cup** *(50 grams)* **shredded Gruyère or Swiss cheese**
2 **tablespoons dry vermouth**
4 **tablespoons butter or margarine**
12 **cooked artichoke hearts (canned, frozen, or fresh)**
2 **cups** *(500 grams)* **cooked and drained mussel meats (canned, or fresh steamed by method on page 31)**
Lemon wedges and chopped parsley for garnish

Elegant Party Entrées

Mussels and Cucumbers in Dill Sauce

Serves 4

1 teaspoon salt
2 cups *(5 deciliters)* **water**
2 medium-sized cucumbers,
peeled, seeded, and cut in
strips
1 tablespoon lemon juice
1 cup *(250 grams)* **cooked and**
drained mussel meats
(canned, or fresh steamed by
method on page 31)
Dill sauce (recipe follows)
Fine dry bread crumbs
About 2 tablespoons butter or
margarine, melted

2 tablespoons butter or marga-
rine
2 tablespoons flour
1 cup *(2½ deciliters)* **milk**
1 egg yolk, beaten
1 teaspoon minced fresh dill
leaves, or ¼ teaspoon dill
seed
Salt and pepper to taste

Add salt to water and bring to boil in saucepan. Drop in cucumber strips and simmer until tender but firm. Drain cucumbers and spread in buttered baking dish. Sprinkle with lemon juice.

Distribute mussels over cucumbers. Mask all with dill sauce. Sprinkle on enough crumbs to cover sauce; drizzle melted butter on crumbs.

Bake uncovered in preheated oven at 400° F *(200° C)* until hot and golden brown.

Dill Sauce

In saucepan, melt butter and blend in flour, stirring until bubbly. Remove from heat; slowly stir in milk. Return to heat and cook, stirring constantly, until thickened. Remove from heat, blend a little hot sauce into egg yolk, then slowly stir egg into sauce. Add dill, salt, and pepper. Makes 1 cup *(2½ deciliters)*; use all.

Index